Conceiving Kinship

Fertility, Reproduction and Sexuality

GENERAL EDITORS:

David Parkin, Director of the Institute of Social and Cultural Anthropology, University of Oxford

Soraya Tremayne, Co-ordinating Director of the Fertility and Reproduction Studies Group and Research Associate at the Institute of Social and Cultural Anthropology, University of Oxford, and a Vice-President of the Royal Anthropological Institute

Volume 1
Managing Reproductive Life: Cross-Cultural Themes in Fertility and Sexuality
Edited by Soraya Tremayne

Volume 2
Modern Babylon? Prostituting Children in Thailand
Heather Montgomery

Volume 3
Reproductive Agency, Medicine and the State: Cultural Transformations in Childbearing
Edited by Maya Unnithan-Kumar

Volume 4
A New Look at Thai AIDS: Perspectives from the Margin
Graham Fordham

Volume 5
Breast Feeding and Sexuality: Behaviour, Beliefs and Taboos among the Gogo Mothers in Tanzania
Mara Mabilia

Volume 6
Ageing without Children: European and Asian Perspectives on Elderly Access to Support Networks
Philip Kreager and Elisabeth Schröder-Butterfill

Volume 7
Nameless Relations: Anonymity, Melanesia and Reproductive Gift Exchange between British Ova Donors and Recipients
Monica Konrad

Volume 8
Population, Reproduction and Fertility in Melanesia
Edited by Stanley J. Ulijaszek

Volume 9
Conceiving Kinship: Assisted Conception, Procreation and Family in Southern Europe
Monica M.E. Bonaccorso

Volume 10
Where There is No Midwife: Birth and Loss in Rural India
Sarah Pinto

Volume 11
Reproductive Disruptions: Gender, Technology and Biopolitics In the New Millennium
Edited Marcia C. Inhorn

Volume 12
Reconceiving the Second Sex: Men, Masculinity, and Reproduction
Edited by Marcia C. Inhorn, Tine Tjørnhøj-Thomsen, Helene Goldberg and Maruska la Cour Mosegaard

Volume 13
Transgressive Sex: Subversion and Control in Erotic Encounters
Edited by Hastings Donnan and Fiona Magowan

Volume 14
European Kinship in the Age of Biotechnology
Edited by Jeanette Edwards and Carles Salazar

Volume 15
Kinship and Beyond: The Genealogical Model Reconsidered
Edited by Sandra Bamford and James Leach

Volume 16
Islam and New Kinship: Reproductive Technology and the Shari'ah in Lebanon
Morgan Clarke

CONCEIVING KINSHIP

Assisted Conception, Procreation
and Family in Southern Europe

Monica M.E. Bonaccorso

Berghahn Books
New York • Oxford

First published in 2009 by
Berghahn Books
www.BerghahnBooks.com
©2009 Monica M.E. Bonaccorso

All rights reserved. Except for the quotation of short passages for the purposes of criticism and review, no part of this book may be reproduced in any form or by any means, electronic or mechanical, including photocopying, recording, or any information storage and retrieval system now known or to be invented, without written permission of the publisher.

Library of Congress Cataloging-in-Publication Data

Bonaccorso, Monica.
  Conceiving kinship : assisted conception, procreation, and family in southern Europe / Monica Bonaccorso.
    p. ; cm. -- (Fertility, reproduction, and sexuality ; v. 9)
  Includes bibliographical references and index.
  ISBN 978-1-84545-112-7 (hardback : alk. paper) -- ISBN 978-1-84545-113-4 (pbk. : alk. paper)
  1. Reproductive technology--Italy. 2. Medical anthropology. I. Title. II. Series.
  [DNLM: 1. Insemination, Artificial, Heterologous--psychology--Italy. 2. Oocyte Donation--psychology--Italy. 3. Family Relations--Italy. 4. Heterosexuality--Italy. 5. Homosexuality--Italy. 6. Social Identification--Italy. WQ 208 B697c 2008]

RG133.5.B66 2008
618.1'7806--dc22
                                    2008034317

British Library Cataloguing in Publication Data
A catalogue record for this book is available from the British Library

Printed on acid-free paper.

ISBN: 978-1-84545-112-7 (hbk), 978-1-84545-113-4 (pbk)

C: 'Mum? I will marry both J. and W' [C's twin boyfriends]
M: 'But you can't...'
C: 'Oh, Yes I can!'
M: 'Especially if you want children, you need to choose one ...'
C: 'No Mum!'
M: 'What do you mean 'No'!'
C: 'K. [at school] told me that she has two dads: one she was born from, the other she lives with'
M: 'So?'
C: 'But Mum, are you a silly billy? [big laugh] My children can have two dads too!'

C.M.R. 5 years old
Conversation held in King's College
Cambridge, December 1999

*Alle mie adorabili bambine,
Camilla Megan e Sofia Margherita*

Contents

Boxes — ix
Foreword — x
Acknowledgements — xiii
Introduction — xv

1. Locating *Conceiving Kinship*: New Subjects, New Boundaries? — 1
   Introduction — 1
   An Overview of Anthropological Enquiry into Assisted Conception — 2
   An Overview of Italian Anthropology — 7

2. Research in Place: Shifting Fields of Enquiry — 15
   Introduction — 15
   Multiple Investigations, Sites, Informants — 17
   Main Investigation — 18
   Comparative Investigation — 25
   Collateral Investigation — 27

3. Heterosexual Couples: Life Plans, Irreversible Infertility and the Choice of a Programme of Gamete Donation — 33
   A Case: Anna and Artificial Insemination by Donor — 33
   Introduction — 34
   Planning Our Life, Planning Our Children — 35
   Discovering Irreversible Infertility — 37
   Choosing a Programme of Gamete Donation — 40
   Normalizing Gamete Donation — 44
   Do It Quickly (and It Lasts Forever) — 47

4. Heterosexual Couples: Gamete Donation, Donors and Biogenetic Make-up — 51
   A Case: Matilde and Egg Donation — 51
   Introduction — 52
   Infertile Couples, Biological Inheritance and Biogenetic Make-up — 53
   Couples' Perception of Donors and Donation — 57
   Good Intentions, Gifts and Donors' Displacement — 62

5. Heterosexual Couples and Clinicians: Strategies in Private Clinics of Assisted Conception — 66
Extract from Field Notes: at Lunch with Clinicians — 66
Introduction — 67
The Provision of Services in Private Clinics of Assisted Conception — 67
Life around Clinics and Clinicians: Trust, Faith and Dependency — 72
The Hyper-*medicalized* Infertile Couple — 75
Managing Recurrent Failure in the Clinic — 77
Getting to Understand Programmes of Gamete Donation — 79
The Work of Kinship in the Clinic — 81

6. Lesbian and Gay Couples Making Families by Donation — 84
A Case: a Lesbian Couple Planning a Family by Donation — 84
Introduction — 85
Lesbian and Gay Couples: Planning a Life Together — 86
Planning Families — 89
Rethinking Motherhood and Fatherhood — 92
The Lesbian and Gay Way: The Procreative Project — 96
The Lesbian and Gay Way: Practices of Inclusion — 99
The Lesbian and Gay Way: Practices of Relatedness — 105

7. The Traffic in Kinship: Southern Europe and Euro-America — 107
Introduction — 107
Ethnographic Reflections: Some Key Notions — 108
Programmes of Gamete Donation: Challenging (in Principle) the 'Model' — 111
Italian versus Euro-American Kinship: Generalizing the 'Model' — 113
A Concluding Note: *Conceiving Kinship* — 115

Appendix I – Assisted Conception in Italy: A Legislative and Political Controversy, 1996–99 — 117
Towards a Unified Text: Political Controversies over Legislation — 118
The Death of the Unified Text: The Rise of a New Controversy — 121
The Political Project Behind Assisted Conception, 1996–99 — 123

Appendix II – Profile of Infertile Heterosexual Couples — 127

Appendix IIa – Profile of Lesbian and Gay Couples — 131

Bibliography — 135

Index — 147

Boxes

1. Note on the Mediterranean Cultural Area Construct 11
2. Note on Catholicism as a Framework 12
3. Note on Social Anthropology in Italy 15
4. The Second Meeting and Beyond 24
5. Note on the Use of Ethnographic Data 31
6. Note on the Ethnographic Data 31
7. Note on the Lack of Statutory Law on Assisted Conception 68
8. The First Italian Law on Assisted Conception, 40/2004 125

Foreword

Conceiving Kinship is set in Italy at a time when the country was regarded internationally as the 'Wild West' of assisted conception. I carried out fieldwork in the late 1990s, when heated parliamentary debates over legislation were taking place (see Appendix I). In fact legislation was only approved years later, in 2004, when the centre-right government of Berlusconi passed one of the most restrictive laws in Europe. It is precisely the long-standing legislative vacuum and political chaos around assisted conception that captured my interest in the first place; it made assisted conception an interesting arena to explore from many different perspectives. At the time, it appeared somehow incongruous that Italy was classified in the anthropological literature as Catholic and Mediterranean (that is, featuring certain distinctive cultural models and practices)[1] and yet that it could leave medical practices as controversial as assisted conception, and in particular those involving third parties' genetic material (i.e. *anonymous* gametes), totally unregulated. It seemed incongruous that the country had left for more than a decade (in comparison with other European countries at the time) such procedures to be the total monopoly of the private sector and had created an American style 'free-market', presenting itself in the international arena as a frontier for *avant-garde* treatment in the field of technological conception, and ultimately a destination for international reproductive tourism. I left for the field intrigued, and somewhat unsure about what I would find.

As soon as I reached Milan, it became self-evident that the political and legislative freedom created a state of chaos as well as a deafening context in which everyone, including my informants, operated. Couples undergoing programmes of gamete donation, clinicians providing services and treatment, institutions, organizations and all those working in the broad field of assisted conception – wished they could just get rid of it all. In looking back at my field notes, and informants' comments on what occurred in the legislative and political[2] arena, it emerges clearly how the chaos was widely felt to be as deeply disturbing. It was unanimously seen

---

1. I will return to this in the book.
2. At the time of fieldwork there were 52 political parties actively operating!

as political madness; somehow the scenario of advances in the field of technological conception raised in the public (including those with vested interests) deep anxieties and apprehension for the future. A sort of metanarrative was articulated, but it felt alienating. Legislators and politicians seemed to constantly play heavy politics; excessively casting contradictory notions of procreation, infertility, family, gender relations, sexuality, biogenetics, identity, personhood and so forth. Politics appeared *inconsistent, reversible and paradoxical*. Politicians seemed to pursue interests and rationales that were often obscure, against the interest of those primarily involved, or even just blatantly against the publics' common sense. An example is the political proposal made at one point to put all frozen embryos up for adoption, the number of which had been estimated at eighty to ninety thousand (see Appendix I). The proposal created (inevitable) furore as, of course, infertile couples and clinicians were outraged, as was the public. What kind of kinship rhetoric would such a suggestion appeal to? Who in Italy at the time would have wanted to make their own embryos openly adoptable, after undertaking invasive infertility treatment? Certainly not the couples I interviewed, who invested many years and often all of their resources in infertility programmes. They were jealously safeguarding their cryopreserved embryos, in the hope of a successful implantation and pregnancy sooner or later. This is just one example; many others could be given. In other words, the chaotic legislative and political climate seemed to fuel metanarratives where all rhetorics - and often *kinship rhetorics* - were up for grabs, as political parties used them to pursue *contingent* and *transient* agendas. To the legislative and political body, it mattered very little (or not at all) that voices and images more often than not appeared idiosyncratic and did not really reflect how the public conceptualise the *kinship implications of technological conception* (this is, interestingly, still the case with the new climate created by the 2004 legislation which has reversed the situation to the opposite extreme).[3] In the same fashion, it mattered little to the legislative and political body that, in contrast to the rest of Europe, Italy was not legislating and was perceived internationally as lawless.

---

3. Interestingly, such rhetorics appear quite problematic for the non-political persona of politicians themselves. Politicians often seemed to need to claim detachment too: if, on the one hand, they were the main actors, on the other hand they periodically called for time out to be allowed to express their views according to personal, and not political, imperatives. This became apparent in the years between 1997 and 1999 when the political game over assisted conception became particularly tough and the political class reached a state of exhaustion. Accordingly, they asked parliament to allow members to deliberate in the light of personal views rather than 'political stands' as it was felt that the continuous manipulation of issues of reproduction, sexuality, life, family, marriage, kin relations and so forth with respect to advances in science, medicine and technology had become unworkable. Politicians had lost the sense of what they were arguing for in the myriad of 'kinship rhetorics' that had been put forward (see Appendix I).

Italy's legislators were more interested in using assisted conception as a battleground for internal politics: as so assisted conception should remain an unlegislated field for several years to come. Italy, as many have already pointed out, is a fascinating political world (see Spotts and Wieser 1986; Martinelli et al. 1999). Once I uncovered the broader political agenda, the apparent contradictions that initially seemed so incongruous to an anthropologist resolved themselves beautifully, and I could devote myself to the study of assisted conception and kinship as the lived experience of my informants. So although the politics is essential context, *Conceiving Kinship* is not about the politics of assisted conception or even about the kinship rhetorics on public display (but see Appendix I for some details). Instead it is an *in-depth journey*, the first of its kind, into how Italian Kinship is conceptualized and constructed by heterosexual couples and lesbian and gay couples making families by donation.

Monica M.E. Bonaccorso
Cambridge, 2006

---

reproduction, sexuality, life, family, marriage, kin relations and so forth with respect to advances in science, medicine and technology had become unworkable Politicians had lost the sense of what they were arguing for in the myriad of 'kinship rhetorics' that had been put forward (see Appendix I).

Acknowledgements

I have collected many debts since I started work on the project that led to this book – I will only be able to acknowledge the most obvious. First of all, I wish to warmly and affectionately thank Marilyn Strathern, my P.h.D. supervisor and mentor, for her continuous intellectual stimulation, encouragement and support since the days of my conversion from Philosophy to Social Anthropology in 1995. There is much that I have learned from her and this is a debt that I will be unable to repay; there is also much I have been unable to learn despite her efforts to teach me. Many thanks also go to Stephen Hugh-Jones who encouraged and helped me from the very beginning of my Cambridge experience to grasp the first rudiments of the discipline, despite my obvious limitations in working with the English language. A special thank you goes to Barbara Bodenhorn and Jeanette Edwards for their insightful comments on the early version of this work, together with the late Sue Benson; they all, in very different ways, gave me the confidence to turn it into a book and I am most grateful for that. I also affectionately wish to thank Paola Filippucci and Nina Hallowell not only for reading the manuscript, but for their care and invaluable support particularly in the later stages of reworking the text. Besides colleagues in the Department of Social Anthropology in Cambridge, including colleagues in the Family Research Centre and the CBA (Comparative Studies in Biotechnology and Accountability), who in various ways have throughout the years offered a stimulating working environment, many thanks also go to Soraya Tremayne and Marion Berghahn for their trust in the project. I am most grateful to my funders in Cambridge, without whom I could not have started the overall venture, part of which is now data for this book: the Cambridge European Trust for a bursary (1996–99), King's College for a scholarship (1996–2000) and particularly Basim Musallam whose support made it possible for me to embark on a Ph.D. at King's. I am also extremely grateful to the Department of Social Anthropology for awarding the William Wyse Bursary (1996–99) and the Ling Roth Scholarship (2000) to complement my funding, as well as for grants and awards obtained to cover my fieldwork expenses: the William Wyse fieldwork grant, the Richard Fund grant and the King's College grant

(1997–98). I also wish to thank the Cambridge Childcare Trust for continuous help throughout the years of my Ph.D. (1995–2000). I must also thank my current employer the University of Cambridge, and the Wellcome Trust, for allowing me in the past few months to work part time on my project 'Cultures of New Genetics'. This has made it possible to revise earlier versions of the manuscript and complete it.

The deepest and most loving thanks go to my exceptionally tolerant daughter Camilla M. who has spent the first five years of her life moving from one place to another, in accordance with my research needs and intricate family circumstances. She recently counted that she moved twelve times whilst I was carrying out this piece of research! Of course loving thanks also goes to little Sofia M. who has just learned to hold a pen and has been enjoying scribbling on the latest versions of the manuscript. Finally, I wish to thank my husband David Rothe – I can now positively answer his late-evening question: 'Have you finished the book yet?'

And finally, I wish to express my profound gratitude to all my informants: couples attending clinics of assisted conception, couples met outside public hospitals, couples adopting children, lay couples, lesbian and gay couples, lesbian and gay activists, clinicians, members of self-help groups and organizations, journalists and policy makers, and last but not least, friends and colleagues who have provided information and various forms of help in facilitating contacts with institutions and organizations, and making it possible for me to carry out fieldwork and write this book. I suspect that not all informants will be entirely pleased with the analysis of the data. Assisted conception, and particularly the practice of gamete donation, for both heterosexual and lesbian and gay couples is an extremely sensitive arena in Italy, and I cannot but apologize if my anthropological eye disappoints the expectations of some or many.

INTRODUCTION

Conceiving Kinship: Assisted Conception, Procreation and Family in Southern Europe

A Journey into Kinship

Conceiving Kinship is set in Italy in the late 1990s. It explores contemporary Italian kinship – notions of procreation, the family, biological and social relatedness – as evoked by assisted conception, and particularly by programmes of gamete donation (that is, programmes that involve anonymous egg and sperm donation). It examines the kinship narratives that surround such programmes in the light of different vested interests, life-styles and worldviews. It compares how heterosexual and lesbian and gay couples making families by donation conceptualize and make sense, culturally, of their choices and experiences. Unable to procreate 'naturally', they both rely on gamete donation to conceive a baby, albeit in different ways: heterosexual couples attend private clinics of assisted conception, and rely on the mediatory role of clinicians and their services for the procurement and use of third parties' eggs and sperm; lesbian and gay couples rely on more informal arrangements outside private clinics, with unknown/known donors and what I have come to call partners in asexual conception. They are part of a fragmented network where transactions and the procurement of gametes take place. In addition, *Conceiving Kinship* explores the clinical context, and the relations between couples and clinicians. The latter play a paramount role in the conceptualization of the programmes as acceptable and viable cultural options (and practices), deploying heavily the language of kinship.[1]

---

1. *Conceiving Kinship* also proposes a collateral investigation amongst three different groups. These are infertile heterosexual couples who are attending public hospitals and are undergoing treatment using their own gametes;

The overall journey aims to explore what assisted conception, and programmes of gamete donation, can tell us about Italian kinship as a cultural form.[2] These particular programmes have been chosen because within the landscape of infertility services they, more than others, throw certain kinship notions into relief. They intrinsically challenge the place that 'Westerners'[3] attribute to biological and social relatedness in the making of kinship. They allow us to examine what it means to be related, to form kin ties, to be biogenetically connected (and disconnected), and to make use of biogenetic anonymous material; ultimately, they allow us to examine what it means to make certain procreative choices that jeopardize culturally established notions of kinship. These are the notions that preferably (in principle) see biogenetic and social (emotional) ties as *exclusive* and *activated in parallel* between the couple in the making of a child, and that see biogenetic ties as non-anonymous. The main endeavour of the book is to trace the kinship 'model' – to follow Strathern's coinage (1992b) – that lies at the core of informants' accounts. What is the model that informants draw upon in making their choices? What constitutes kinship at this point in time? What kinship notions keep recurring in the narratives? How are they deployed, exploited and reformulated in the context of medicalized forms of conception – particularly when the genetic material of third parties aids the project of a baby and starting a family? How do informants reconceptualize their biological and social ties in view of such contribution? How do they make sense of their choices when, as they consistently claim, they are looking for normalness, ordinariness and a life like everyone else? What does it mean to long for a 'life like everyone else'? Finally, do the different narratives of lesbian and gay couples illuminate the kinship notions at play? Do lesbian and gay couples deploy a divergent kinship model or the same?

The answers to these questions are far more complex than expected; and they themselves generate a new set of questions, of an altogether different nature. If, on the one hand, the journey into assisted conception and programmes of gamete donation makes *Italian kinship* fully explicit, on the other, the very kinship model that emerges poses questions with wider implications for anthropological imaginings of European boundaries, cultural continuities and discontinuities. To paraphrase Anderson (1991),

---

infertile heterosexual couples who equally suffer from infertility but instead have decided to adopt; and couples who have no vested interest in assisted conception but who are willing to express their views on such medical practices. These additional investigations offer contrasting data for further reflection. These views will be found in footnotes so as not to break the flow of the core ethnography.
2. This work is thus not solely an account of assisted conception and programmes of gamete donation per se – although this is inevitably part of the project and a necessary endeavour given the scarcity of ethnographic data in the field from Southern Europe.
3. I beg the reader to suspend judgement on the broad use of 'Westerners'. I'll return to this at a later point in the work.

there are suddenly more reasons to imagine 'imagined communities' differently – or to reimagine them in a different light. The kinship model that emerges through the persistent voice of informants is, surprisingly, not so exclusively Italian and Mediterranean. The *mediterraneità* (mediterraneity) that one would expect to find, that reflects certain particular features, and that would allow the data to be pigeonholed as Mediterranean in essence, is not fully apparent. In contrast to the anthropological literature on Italy, and on Italian kinship, from the famous and much criticized Banfield (1958) to more recent works (see Chapter 1), the data emerging from *Conceiving Kinship* fundamentally question strict definitions. Abu-Lughod's (1991) warning is thus fully confirmed: culture(s) are not bounded and discrete; they are not isolated units. The ethnographic material shows that informants shift the boundaries of kinship in fascinating, at times contradictory and ambivalent ways, drawing upon notions that anthropologists are familiar with in contexts such as Northern-Europe and America (in short, Euro-America; see Strathern 2005: 163). The material calls for continuous reference to the Euro-American literature on assisted conception and notions of kinship that emerge there (see the work of Price 1992, 1995; Ragoné 1994; Dolgin 1997; Franklin 1997; Daniels and Haimes 1998; Edwards et. al. 1999 [1993]; Franklin and Ragoné 1998; Davis-Floyd and Dumit 1998; Becker 2000; Edwards 2000) as well as to the work of Strathern on 'English' kinship (1992a, 1992b) and Schneider on 'American' kinship (1980 [1968]). In other words, what emerges is a proximity with an unexpected literature and a kinship model that Strathern has powerfully called Euro-American, but has circumscribed and confined deliberately so as not to be Southern European. She writes:

> Now while I take 'English' as my exemplar of a folk model and thus illustrative of Euro-American kinship thinking, there is also good reason to suppose that the trivialisation of kinship in social life is a characteristic that may well distinguish it from some continental or Southern European models (though it may give it affinity to aspects of 'American' kinship). It is of interest insofar as it has helped shape British anthropological theorising on kinship. Both belong to a cultural area I have called 'modernist' or 'pluralist'. (1992b: 106)

This of course leaves us with a number of problems, and very few solutions. What makes the similarity, what makes the difference? If culture is never free-floating, decontextualized, entirely 'deterritorialized' as Xavier Inda and Rosaldo argue: 'cultural flows do not just float ethereally across the globe but are always reinscribed ... in specific cultural environments' (2002: 11) what makes Italian kinship Italian? Conversely, what makes Euro-American kinship Euro-American? Moreover, how far can we generalize? Is it *Conceiving Kinship*, and the focus on assisted conception in particular, that sheds light on the non-exclusiveness of certain notions or is it the previous focus of Italian kinship studies that may have obfuscated it? In other words, is assisted conception illuminating a pre-existing kinship model which earlier ethnographic and analytical approaches to the area did not and could not reveal (see Chapter 1) or is it shedding light on an emergent kinship model. Is it thus the case that, to paraphrase Cambrosio, Young and Lock (2000), biomedical objects and subjects call for new kinds of analysis? Do they operate as an analytical window making explicit the implicit, as Edwards points out (2000),[4] or do they do much more than that? Do they perhaps reshape and redefine notions of kinship to such an extent that they shorten *cultural distance*? The latter hypothesis is difficult to endorse as it contrasts with increasing ethnographic evidence from 'non-Western' locales – where it emerges that advances in the field of assisted conception *mould around* established kinship models and practices in complex, articulated and fascinating ways (see Kahn 2000; Inhorn and Van Balen 2002; Inhorn 2003, 2005; Bharadwaj 2000, 2003). Such a hypothesis, of course, would also raise questions about what Hannerz (1996) calls the 'opaque' relationship between the global and the local (see also Appadurai 1996). It is not the aim of *Conceiving Kinship* to fully answer all questions, but to pose some questions for European kinship studies and for comparative anthropology as these powerfully surface from the ethnographic data. The next chapter will discuss in more detail some of the relevant literature on assisted conception and some of the Italian literature on kinship that bears on these issues, with the intent of setting out the context and not of resolving the dilemma.

Conceiving Kinship in Detail

Conceiving Kinship differs from other studies in the area of assisted conception. First of all, it is the first anthropological and ethnographic account of assisted conception located in Italy. It is set at a time when private clinics were unregulated and operated with complete autonomy and freedom in a legislative vacuum. This gives uniqueness to the data in that the kinship narratives are not clouded or constrained by regulatory frameworks. Secondly, at the time of writing, it is still one of very few works located in Southern Europe. As the overview of the literature in chapter 1 will show, apart from a few sporadic cases, Southern Europe has been mostly neglected and Northern European and American perspectives and focus have overwhelmingly dominated the overall field. Only very recently new investigations in 'non-western' locales have started to take place. Thirdly, *Conceiving Kinship* offers a new multi-sited (Marcus 1995; 1998) and comparative perspective (Gingrich and Fox 2002): it simultaneously explores different sites and compares different

---

4. Edwards writes: 'NRT problematize taken-for-granted premises and, in this sense, they act as an ethnographic window through which notions of what constitutes relatedness, which usually remain implicit, can be discerned' (2000: 31).

experiences of gamete donation. It also adds the perspectives of those who object to gamete donation. Finally, it offers a comparative reflection with the kinship literature produced in the so-called Euro-American geographical area/conceptual domain.

Chapter 1 locates *Conceiving Kinship* within two bodies of literature: the anthropological literature on assisted conception, and that on Southern European kinship. The first section presents an overview of works on assisted conception located in Britain and America, it highlights their concentration (in the English-speaking world) in the mid-to late 1990s, and points to a lack of equivalent focus on Southern Europe. The second part of the chapter overviews the Anglo-American anthropological literature on Italian kinship – addressing some of its limits due to a prevalent and uninterrupted interest in rural communities, mostly in the south of Italy, and an overall stronger tradition in the field of politics, ideology and religion rather than kinship. The chapter concludes with two boxes: the first, 'Note on the Mediterranean Cultural Area Construct', raises questions about prevailing anthropological imaginings of the area and the boundaries created by theoretical and methodological scholarly traditions. The second box, 'Note on Catholicism as a Framework', illustrates certain distinctive features of Italian Catholicism and the context in which it operates. It points out why a Catholic framework cannot be taken too much for granted. Overall this chapter is informative for those who are not familiar with the anthropology of assisted conception and of Italian Anthropology too.

Chapter 2 describes the investigation, the research setting and the social relations that made fieldwork possible amongst clinicians operating in private clinics of assisted conception, infertile heterosexual couples suffering from impaired infertility and undergoing programmes of gamete donation, infertile couples undergoing programmes with their own gametes, couples with no vested interest in assisted conception, couples suffering from infertility and attending programmes with their own gametes, adoptive couples and, finally, lesbian and gay couples. It also contextualizes the research choices and the methodological approach, including what I have called the 'without-method approach'. This is a strategic device that has made it possible to get access to clinics of assisted conception at a time in which, due to the legislative vacuum, clinicians were protective about their practices and often unwilling to easily grant access to a scholar working within an unfamiliar disciplinary field. This chapter, by offering a full account of the investigation, is part of the ethnography itself.

Chapters 3 and 4 are dedicated to infertile heterosexual couples suffering from impaired infertility and seeking treatment with donated gametes. These are couples, who, having failed on programmes with their own gametes (in the majority of cases having done so in public hospitals), have moved to the private sector to undergo a programme of gamete donation. Chapter 3 portrays their state of mind, retrospectively and in the light of the diagnosis of irreversible infertility. Couples talk about their plans and expectations in life, their ideas – and ideals – of a family with children. They also address their feelings and emotions, the sense of loss, damage, and limitation they now experience given the diagnosis and the choice they have made to undergo a programme of gamete donation. Between making that choice and attempting to normalize it they reveal what it entails to rely on anonymous egg and sperm donors, yet to claim to be looking for normalness, ordinariness and a life like everyone else. They reveal what it takes to come to terms with a choice that is experienced as fundamentally contradictory and ambivalent. The contrast between the dream of a family with their own children, where children are genetically related to each parent and resemble both parents (as an ideal family configuration) and the reality of a programme where children are genetically related to one parent only and the anonymous donor, emerges powerfully.

Chapter 4 unfolds the case further. It explores couples' conceptualization of biological inheritance and biogenetic make-up (*eredità biologica* and *patrimonio genetico*), perceptions of donors and the practice of donation, and related ambivalence. It shows how complex it can be to undergo programmes of gamete donation with third-party assistance, and then conceptualise and sustain the choice in everyday life and in the long term. If, on the one hand, such programmes allow couples to conform and perform certain cultural and social prescriptions (having a child, making a family), on the other hand, they profoundly jeopardize other norms (correspondence between biogenetic and social relatedness). With a programme of gamete donation couples need to revisit the notion that 'blood is thicker than water' (see Schneider 1980 [1968]; Wolfram, 1987) – although in asymmetrical ways. Blood *is* thicker than water with reference to the biological tie that is preserved between the fertile parent and the future baby, but *it is not* with reference to the *anonymous* donor and the baby. This is a difficult exercise – particularly for couples looking for normativity.

Chapter 5 points out how the lack of legislation has emphasized a specific rhetoric about treatment, services and choice. It examines the interactions between providers of treatment and couples. Programmes of gamete donation are designed, applied, managed and sold by private clinicians. In selling programmes, clinicians also sell ways of thinking about them: ideas are pulled together from different domains, the medical language is replaced with what I have called the 'language of commonplaces': accessible and familiar to anyone approaching treatment, able to cut across class and status, and touch to the core of very specific kinship notions. It is a language that suppresses anxieties, fears, doubts, feelings of incongruity. It keeps couples going and able to deal with repeated failure. Interestingly, this is the same language used by clinicians to rationalize their own role as medics. Clinicians need to believe in the technologies too.

Chapter 6 presents the comparative ethnography with lesbian and gay couples making families by donation. It offers a lesbian and gay account of the themes treated in Chapters 3 and 4, although necessarily in a different form. It is divided into two parts: the first discusses lesbian and gay partnership, life plans, the desire for a child and lesbian and gay ways to arrange conception with other couples or individuals. The second part discusses gamete donation, attitudes towards biogenetics and biological ties, the role of unknown donors, known donors and what I have called 'asexual partners in conception'.

The conclusion, Chapter 7, highlights some key notions emerging from the ethnographic journey (including those arising from couples without vested interest, infertile couples not choosing programmes of gamete donation and/or opting for adoption, found in the footnotes) and sets them side by side with some key notions, mostly framed by Strathern (1992a, 1992b; see also 2005) and Schneider (1980 [1968]). This exercise highlights interesting continuities between Italian and Euro-American kinship. The subsequent sections address the challenges that programmes of gamete donation pose as well as the limits of generalization. The final section concludes with a note on *Conceiving Kinship* as a whole.

These concluding remarks question long-standing certainties within the discipline of social anthropology and encourage new thinking about some conventional ways to 'imagine' European kinship and its boundaries. Although this is not a cross-cultural ethnography per se, the reference throughout the ethnography to Northern-European (mostly British) and American works allows for a sort of conceptual cross-cultural comparison. As Carsten has pointed out, 'a century or more of cross cultural comparison of institutions of kinship has taught anthropologists to take little for granted in the way people live out and *articulate notions of kinship*' (2004: 17, emphasis added). Whilst Italian kinship has always been framed as characteristically Mediterranean, the ethnographic material confers with Carsten's view. *Conceiving Kinship* does not intend to offer definite answers; that would be a life-time project. It only intends to pose some questions, as a result of an ethnographic enquiry into kinship at the edge of medical advances in Italy. By pushing certain conventional boundaries the usually unsaid is said and the inexplicit is made explicit: Italian kinship is revealed under a different light.

Note on Writing

Conceiving Kinship has been written making extensive use of quotations so as to take the reader through the ethnographic journey I undertook. These should help to put in greater relief those kinship notions that soon become recurrent and that seem to form a consistent pattern, albeit sometimes contradictory. The voices of informants soon form a *Weltanschauung*. However, the reader should be aware that I have made a heavy selection of data – I have omitted to include much material in order to avoid constructing multiple texts with numerous threads that I would thereafter be unable to follow; I have accordingly written a text within many possible texts. This is a common, although in past years highly debated, practice amongst anthropologists (see Clifford and Marcus 1986; Marcus and Fisher 1986). Besides the obvious reasons that I could not have written about everything, some data has been omitted as it is too sensitive: for instance, interviews with donors, as these may have upset many couples. Some other omissions have been strategic: they would have distracted attention from the overall purpose. Two omissions are particularly significant: first of all, I make no analytical distinction between egg and sperm donation. I leave unexplored whether couples talk about them differently, if they make a distinction between using egg and sperm donors, if they see the two practices as having the same or different implications and if husband and wife produce different accounts. Although I have collected material that would be informative in this respect, and couples have produced at times different gender statements depending on whether they were undergoing egg or sperm donation, I have decided to leave it out in this context as it is not relevant to the crux of the analysis. This is because *Conceiving Kinship* does not want to focus on gender differences within the heterosexual couple as such, but on core notions of kinship as evoked by egg and sperm donation. Secondly, I make no distinction between conceptualizations and choices made by lesbian couples and by gay couples. Of course, there are some interesting and powerful differences in their ways of thinking, planning and conceptualizing families by donation. It may be surprising that I do not linger on those; I almost gloss over and treat them with a certain degree of homogeneity. Once again, this is a deliberate choice and there is no intention to dismiss the different politics and other aspects that influence lesbian and gay families by donation (although this is a complex area and the 'differences' that are popularly addressed are not the differences that I would want to highlight, see Bonaccorso 1994). The lesbian and gay narrative is meant to be comparative, so as to highlight certain kinship notions. It is not meant to be a self-standing account. To draw a sub-distinction, once again, would have distracted attention from the crux of the comparison.

On a different note, the ethnography is written in the ethnographic present, 'the practice of giving accounts of other cultures and society in the present tense' (Fabian 1983: 80), although in some cases this may appear inappropriate as I refer to programmes of gamete donation which have become recently unavailable. Finally, a better translation of the quotes has been made since they appeared for the first time in my Ph.D. 'The Traffic in Kinship: Assisted Conception for Heterosexual and Lesbian-Gay Couples in Italy' and the names of couples have been further disguised to avoid the quotes becoming attributable. I have also omitted to name hospitals, self-help groups, private clinics and other informants unless they were already public, such as in the case of politicians working on legislation (Appendix I). This sometimes injects a sense of vagueness into the narrative, which contrasts with the vividness of the quotes, but of course to identify places and people would not necessarily add depth to the overall account.

CHAPTER 1

LOCATING *CONCEIVING KINSHIP*: NEW SUBJECTS, NEW BOUNDARIES?

Introduction

This chapter aims to locate *Conceiving Kinship* within two bodies of literature: the anthropological literature on assisted conception, and that on Italian Anthropology. The first section offers an overview of works located in Britain and America, pointing to those that will be most informative here. Inevitably, such an exercise highlights the overwhelming predominance (in the English-speaking world) of studies informed by a so-called Euro-American perspective (as several authors themselves claim) and the lack of an equivalent anthropological focus on Southern Europe. It also highlights a concentration of works, within the disciplinary field of social anthropology, in the mid to late 1990s. The review also points to some examples on 'non-Western' locales. The second part of the chapter presents an overview of the Anglo-American anthropological literature on Italian kinship – pointing out some of its limits due to a tendency (for some time) to focus on rural communities, mostly in the south of Italy, and more generally a stronger tradition in the field of politics, ideology and religion. Finally, the chapter concludes with two boxes: the first, 'Note on the Mediterranean Cultural Area Construct', introduces the debate and raises questions about prevailing anthropological imaginings of the area and the cultural boundaries created by theoretical and methodological scholarly traditions. The second box, 'Note on Catholicism as a Framework', highlights certain distinctive features of Italian Catholicism and the context in which it operates. The significance of Catholicism cannot be taken too much for granted; it needs to be contextualised so as to avoid simplistic generalizations. This chapter is informative for those who are not familiar with the anthropology of assisted conception and of Italy. Overall, it will help to contextualize better the discussion that will follow on certain similarities/continuities between Euro-American and Italian kinship.

An Overview of Anthropological Enquiry into Assisted Conception

Northern Europe and America

The first British anthropological works[1] on assisted conception in the UK appear as commentaries on the work of the Committee of Enquiry into Human Fertilisation and Embryology set up in 1982 by the UK government 'to examine the social, ethical and legal implications of recent and potential developments in the field of human assisted reproduction' (1985: vii). A pamphlet authored by Dame Mary Warnock, who chaired the committee, was published with the title A *Question of Life*. A few years later, in 1989, another document, *Fertility and the Family* (commonly known as *the Glover Report*), was also published and is often mentioned in the work of British social anthropologists. Peter Rivière describes the Warnock report as 'a fascinating document ... deserving of some anthropological commentary' (1985: 2). Fenella Cannell describes it as 'a source of provocative insights into English constructs at a moment of crises posed by the new reproductive technologies'[2] (1990: 668). Chris Shore, two years later, also analyses both the Warnock Report and the Human Fertilisation and Embryology Bill of 1990 in the light of the interest that society has in maintaining symbolic power over issues of fertility and reproduction. He draws attention to the so-called virgin-birth scandal, which appeared in the press, as an example of the controversy that developments in assisted conception increasingly create (Shaw and Borlass 1991; Dyer 1991, cited in Shore 1992).[3] In the same period Marilyn Strathern, the social anthropologist who since the early days has been the most prolific on the subject, explores questions of consumer choice (1990), gender models (1991), surrogacy and notions of new/old family forms and the NRTs (1998). In *Reproducing the Future* (1992b) she discusses advances in the field of technological conception drawing on

---

1. This brief overview aims to provide some background for those who are not familiar with the anthropological literature on assisted conception. The chronological order should help to illustrate how the focus of these studies has shifted over time. I am limiting this overview to works which are written in the English language, and that seem of most relevance here.
2. I will continue to use the construct 'new reproductive technologies' (NRTs) when referring to the period in which they have been widely termed so. More recently the language has shifted to 'assisted conception'.
3. In 1991 the case of a 20-year-old woman who had never before had sex and was undertaking artificial insemination in a clinic in Birmingham appeared in the press. The case led to a hot controversy on whether single women (especially lesbians) should to be allowed to have children by artificial insemination.

both Euro-American kinship and contrasting Melanesian material, delving into conceptualizations of personhood, body and generation. She also contributes to Edwards and colleagues' volume *Technologies of Procreation* (1999 [1993]). Overall the volume tackles the most pressing issues concerning technological conception from an entirely anthropological and ethnographic perspective, and will become a point of departure for further work carried out by other scholars. There, Strathern reminds the reader that, of course it is one thing to say (as it is repeatedly said) that such technologies have all sorts of implications 'foreseen and unforeseen for society. But how one demonstrates the significance of those implications is another matter' (1993: 11).[4] The papers that follow are set to demonstrate precisely that. They investigate prevailing implications raised by the NRTs from different focal points: Jeanette Edwards presents an ethnographic enquiry into lay views (from the northwest of England) on the various possibilities offered by the technologies – she investigates how these are conceptualized by informants using their knowledge and experience of kinship (see also subsequent work where Edwards returns to the same village and investigates in further depth her informants' views on developments in the field of assisted conception – see especially 2000). Eric Hirsh's interviews, again with lay informants, are carried out in southeast England and touch on what it means to marry, live together and have a family. Informants delve into the issues in the light of the NRTs and other possible configurations, such as those created by adoption. Sarah Franklin proposes an analysis of parliamentary debates over the regulation and the passage into law of the Human Fertilisation and Embryology Bill introduced in 1989 and enacted in 1990. Francis Price explores clinical practices in the context of the NRTs. At around the same time two other collected volumes are published: McNeil and colleagues' *The New Reproductive Technologies* (1990) and Meg Stacey's *Changing Human Reproduction* (1992; see also Stanworth 1987). They play a similar role, although contributions are more widely from the social sciences. They tackle fully the impact of the NRTs on women's lives – as McNeil puts it, 'we never just purchase a piece of technology – we are buying a transformed life' (1990: 2). As such, technology can always 'turn' against women (Petchesky 1986, cited in McNeil). Both these works have their roots in groundbreaking feminist accounts in the field (see Corea 1985; Corea et al. 1987; Stanworth 1987; Spallone and Steinberg 1987; Spallone 1989; see also Franklin and McNeil 1988; Arditti et al. 1989). Challenges to the notion of the family (1990) and the social management of genetic origin (1992) in the context of gamete donation are extensively and

---

4. This is a very relevant point that should not be taken too much for granted. A review of the literature clearly shows that since the late 1980s/early 1990s scholars across several disciplinary fields have often taken such implications for granted. It is often stated that the technologies come with all sorts of implications for culture and society, but it is also very often left unsaid what those specific implications are.

notoriously addressed in the UK by Erica Haimes. Haimes also explores how gender is deployed in practices of gamete donation and, as she puts it, the politics of 'equivalence between egg and sperm donation which in truth underlies unstated assumption about their difference' (1993: 85). In 1998, Haimes with Daniels, put together a collection from the social sciences with the title *Donor Insemination* to offer a 'multidisciplinary [analysis], involving the combination of the "sociological imagination" with "historical sensibility" and "anthropological insight"' (1998: 3). In this volume the authors locate the practice of donor insemination (DI) in a more international context and illuminate 'the complex web of social relationships that DI both reflects and constitutes as part of a wider social order' (Daniels and Haimes 1998: 1). The work of Francis Price (in addition to the one mentioned above) also focuses on gamete donation – donor matching, anonymity and secrecy and what she significantly calls 'the culture of concealment' that existed particularly at the time of her writing (see 1992, 1995, 1996 – legislation has recently changed in the UK). Price's work on gamete donation, like Haimes', is of special interest here, as they both examine gamete donation in depth, from different but complementary perspectives. A salient ethnography of couples (and women) who have undergone IVF treatment in clinics of assisted conception in the UK is that of Franklin (1997). The study explores couples' perceptions of risk and success of IVF programmes, involvement with technology and certain expectations that are inevitably produced; it highlights IVF treatment as an experience that 'takes over' and becomes a 'way of life', and the constant emotional and physical work that goes into it. It also powerfully brings to light couples' common feelings that they 'have to try', as well as degrees of ambivalence when trying means also exposing oneself to recurrent failure. The availability of technology itself pushes couples to explore all possible options. Konrad's work (1998, see also 2005) focuses on non-IVF egg donors and IVF recipients, in clinics of assisted conception in Great Britain. It explores, what she calls, 'nameless and untraceable relations' between those involved both in the donation and in the reception of gametes. Gametes here play a double role: they invisibly form relations, but also deny them as they are passed from stranger to stranger with the help of mediators (clinicians and clinics). They reach their destination anonymously.

On the other side of the Atlantic several anthropologists have been equally fascinated by the advent of the NRTs. Due to the free market in which infertility services are made available to couples able to afford them, even more controversial issues emerge. In 1994, Helena Ragoné writes a fascinating anthropological account of 'closed' and 'open' surrogacy programmes in the United States which revisits consolidated notions of motherhood, fatherhood, the family and kinship (see also 1996). In the same period, anthropologist and lawyer Janet Dolgin focuses on the status of surrogacy in the United States and on contentious court cases (1990). In *Defining the Family: Law, Technology and Reproduction* (1997), she explores more widely the implications of advances in the field and the law. Dolgin shows how the two do not progress together, and how the rhetoric of the family based on biological ties is increasingly in the way She powerfully points out how courts seem at times to have difficulties in facing questions about the significance of biological versus social relatedness. She delves into questions such as who is the mother in surrogacy agreements – the gestational or the genetic contributor, or instead the commissioning one who is longing for a child? At the time of her writing these are unresolved questions, not just from a kinship perspective, but from the perspective of the law. The papers of Franklin, Ragoné and Cussins, in Franklin and Ragoné's volume (1998), deal with the cultural strategies used to normalize and naturalize both technological conception, infertility clinics' procedures and, ultimately, couples' choices from different angles. These again are illuminating works because such strategies are salient to the success of infertility programmes. Becker's (2000) work is again situated in the United States. It examines the experience of infertility, couples' choice to undergo treatment together with notions of biological and social parenthood, questions of normalcy, cultural ideologies and ideals and how these are mediated with 'the realities of people's lives' (2000: 3). Becker's work also pays attention to gender dynamics in the context of infertility and treatment. Charis Thompson's work (2005), equally located in the United States, also focuses on certain gender prescriptions as played out in the context of infertility treatment. Thompson examines, for instance, how notions of masculinity are triggered by the stigma that is often attached to infertility and the role that partners perform during treatment. The work focuses on processes of normalization and routinization during treatment, on the re-elaboration of kinship and ethnicity and, finally, on couples' agency.

Europe and Beyond

Outside Northern Europe and America, from the mid-1980s and beginning of 1990s, other scholars work on similar issues. In France, for instance, Simone Bateman Novaes as early as 1986 (see also 1989, 1998) works on semen banking, artificial insemination by donor, medical discourses surrounding the practice of donation and issues of anonymity. Bateman Novaes's work is particularly illustrative as it brings to light the social and cultural meaning attached to the donation of gametes from several perspectives. In Spain Stolke examines the cultural meaning of infertility in terms of both 'freely expressed desires and socially induced needs' (1988: 11). Stolke also examines how the 'NRTs – fertilisation with donated semen, ova or embryos – ... [seem to] challenge conventional biological concepts of parenthood' (ibid.: 12), and hence might erode and transform traditional concepts of marriage, family, filiation and inheritance. She then concludes how this is not the case and, on the contrary, argues for the tendency to 'regulate the effects of the NRTs in terms of established father-centred institutions and norms' (ibid.: 14). In her view, the crucial question remains fatherhood: the NRTs, and indeed artificial insemination by donor, recreate the same 'old-fatherhood', and not alternative forms of it. More recent work in Spain on egg donors explores the meaning that women seem to attribute to gamete donation and how they articulate notions of kinship (Orobitg and Salazar 2005). In Italy, to date, most work appears in collateral disciplines such as psychology (see Vegetti Finzi 1998), journalism, gender/feminist studies (see for instance Valentini 2005; Di Pietro and Tavella 2006), bioethics/legal studies (Franco 2005). Chiara Valentini's work (2005) consists of a series of journalistic-styled interviews with infertile couples who have undergone infertility treatment; the work underlies how couples' suffering has been alleviated by treatment, and by successful outcomes - the so-called 'taken-home baby'. It is set to demonstrate how the new law (40/2004) often severely constrains couples in achieving their desire for a baby and a family. In contrast, Alessandra Di Pietro and Paola Tavella (2006) take a strong feminist stance against all developments in the field of assisted conception (and widely new science and new medical technologies) that claim to improve human life from a very early stage. The authors not only confute the new medical technology thesis, but argue that such advances viciously limit human - and women's - freedom by shaping particular forms of society. Their work is important as an exemplar of a radical feminist Italian tradition (see Bono and Kemp for a short history of Italian feminism, 1991). An in-depth bioethical and legal analysis of Italian assisted conception, in the light of the new 2004 restrictive law, is offered by Vittoria Franco (2005), a philosopher and politician herself. [5]

The work of Susan Kahn (2000) and Marcia Inhorn (2003) are examples of new departures and explorations in the field in 'non-Western' locales. Kahn's work, *Reproducing Jews,* is set in Israel and focuses on contemporary rabbinic attitudes to the new reproductive technologies, and their use by women (mostly single women) in line with religious and cultural prescriptions. Kanh's work describes how such technologies are debated within Jewish law and came to be used in medical settings – and what are the implications for conception of kinship, and specifically the re-elaboration of both biological and social ties. Inhorn's study (2003, see also 2006) is set in Egypt and focuses upon different couples using the NRTs. The study explores the stigma attached to childlessness as well as infertility treatment, in view of the fact that neither gamete donation nor alternatives such as adoption are viable options within the Islamic context.

Other examples of investigations in 'non-Western' locales, with contributions from anthropology, are contained in the volume *Infertility Around the Globe* (2002), edited by Inhorn and Van Balen. Here scholars set out to investigate, with studies that differ in nature, what they call 'non-normative reproductive scenarios and experiences' (2002: 4) in a variety of cultural contexts. The work of Lisa Handwerker, set in China (2002)

---

5. With the exception of a few Italian works, all the works cited here are in the English language.

addresses the ambivalence of the NRTs in contexts where they are used to overcome infertility but also as tools to promote notions of modernity and 'new eugenics': the NRTs feed the ideal of the perfect one single child (male), setting new standards of normality in the minds of both professionals and parents-to-be. Bharadwaj's work in India (2002) illuminates the media landscape of infertility treatment and the credit that media appearances confer to medical professionals; further work (2003) sheds light on couples' choice of treatment with gamete donation as opposed to adoption; gamete donation can be kept secret whilst adoption exposes the infertility problem and the attached stigma.

This brief historical/geographical account of some of the most significant works in the broad field of assisted conception highlights, on the one hand, the shift of interest from the new reproductive technologies as a broad arena of technological application to infertility and conception, to more specific aspects arising from the use of third parties' genetic material as well as surrogates; on the other hand, it shows the great – and continuous – prevalence of works located in Northern Europe and America and the increasing familiarization with the idea of Euro-America as a notion and as a domain. Further, it shows the more recent trend in cross-cultural work in the field of assisted conception. The latter, besides being informative in its own right, if read with a comparative eye, shows how the reproductive technologies, the form in which they are offered by clinics and clinicians and 'bought' by those unable to conceive, always mould around existing social, cultural and religious practices. Overall, it thus emerges that in all cultural contexts clinicians and couples (as well as the state) conceptualize technological treatment in ways that fit with existing cultural and social prescriptions and practices; the medical and reproductive technologies are spreading cross-culturally, but they lack the power to fully upset pre-existing cultural forms. Their occurrence is fascinating precisely because they help reinstate the cultural norm, although via means that are often unconventional.

An Overview of Italian Anthropology

Kinship Studies

In Anglo-American social anthropology, there is not a long and rich tradition of kinship (and gender) studies in Italy, in contrast to other parts of Southern Europe.[6] The anthropological interest in Italy, which started

---

6. The literature on kinship and gender in Southern Europe, in the English language, is more substantial than in Italy. For Greece see for instance the work of Du Boulay (1974, 1984), Herzfeld (1983, 1987), Dubisch (1986), Loizos and Papataxiarchis (1991), Cowan (1991), Just (2000), Saint-Cassia (2006); for Malta see Mitchell (2002); for Spain see the work of Brandes (1980), Bestard-Camps (1991), Collier (1997), Brogger and Gilmore (1997), Gay-y-Blasco (1999); for Portugal that of de Pina-Cabral (1986), Cole (1991); and for France the work of Segalen (1991) and Rogers (1991).

in the 1960s and developed throughout the 1970s to the late 1990s, shows a focus on predominantly small peasant communities in the south of the country, in impoverished and economically underdeveloped areas. Such an approach, particularly in the early days, corresponded to the tendency of British anthropologists to constitute the Mediterranean as a new, yet still exotic, area of anthropological enquiry (see Llobera 1986; de Pina-Cabral 1989), and complied with American anthropologists' interest in exploring homelands of migrants to the United States (see de Pina-Cabral 1989). Signorelli, an Italian anthropologist, points out how the prevailing peasant village approach has been particularly central in the Italian case; she also emphasizes the need to carry out urban studies and critiques the overall trend in the Anglo-Saxon approach:

> A first observation ... although Italy is the European country that can claim the oldest and most established network of cities, and although Italian culture traditionally valorises urban milieux against rural ones (Silverman, 1986), there are very few anthropological works on Italian cities, by either Italian or foreign scholars ... The passion of Anglo-Saxon researchers for peasant villages, for objects of research decontextualized from historical, geographical and political contexts, can be explained with disciplinary interests (Saunders, 1995). Recently, however, the hypothesis that the construction of this image of Italy (as well as of other Mediterranean countries) had political agendas has been suggested (Hauschild, 1995). (1996: 16, my translation)

The first anthropological study on family and kinship in Italy is that of Banfield (1958), located in the Lucanian community of Montegrano, in Basilicata, in the south. Banfield addresses, what he calls, peasants' 'backwardness' with his famously criticized notion of 'amoral familism' by which inhabitants would not cooperate with one another outside the boundaries of their immediate families. While this study has been criticized (see for instance Miller 1974), later studies, although ethnographically richer and more historically and socio-economically contextualised, tend to retain a rural focus – see for instance the work of Davis (1973), Bell (1979), Esposito (1989) and Galt (1991). These works all share a focus on rural communities. Davis' work in Pisticci, in the southern Italian region of Basilicata, focuses on a peasant community, land tenure and the relationship with family and kinship; it examines the intricacies between land ownership and kinship. Bell's work is based on four communities between the Emilian Apennines and inland Sicily, and examines fate, honour, family and village – the values that are believed to be the pillars of rural culture. The work of Galt in Locorotondo, Apulia, in the south of Italy, again focuses on land ownership, residence, family structure and social values. The first shift to an urban setting, although still in the south of Italy, occurs with the study of Belmonte (1979), set in the neighbourhood of Naples, amongst the sub-proletarian poor. Besides strategies of economic endurance, the study focuses on family life and kin ties as mechanisms of survival. Goddard's study (1996) is also located in urban Naples, again in the south, and examines more closely women involved in employment and the effect on family and gender relations. Kertzer and Saller's volume (1991) is a joint effort between historians and anthropologists; it tackles notions of change and continuity in Italian kinship, exploring why and how family life has changed over a two-thousand-year period. A radical move to the north of Italy, in the region of Como in Lombardy and in a fully industrialized setting, is made for the first time by Yanagisako who explores gender and kin relations in family businesses (1991, 2002).[7] The work of Yanagisako is important as it constitutes a departure from previous interests and focus; the overall investigation sets a new direction for anthropological studies on family, gender and kinship in Italy. The work of Sciama (2003) is also located in the north, on the Venetian Island of Burano, and examines identity, gender, honour and kinship as they emerge in the inhabitants' love for the environment and the skills of fishing and lace making. A recent work, in the urban and northern setting of the town of Parma, is that of Sonja Plesset (2006), which focuses on intimate-partner violence, family and gender relations. It examines the relationship between Italian feminism and its impact in women's daily lives, tradition and modernity. Other anthropological works of anglophone scholars (and more rarely Italian anthropologists trained in the Anglo-American tradition) located in Italy to date touch on kinship (and/or gender) only tangentially; they have as their main focus patronage, religion, politics, economics, ethnicity and localism (see for instance Cole and Wolf 1974; Schneider and Schneider 1976; White 1980; Kertzer 1980, 1996; Douglass 1984; Pardo 1996, 2000; Stacul 2003), and notably the mafia (Blok 1975; Arlacchi 1983; Gambetta 1996). Such a predominance and focus on subjects other than kinship and gender partly reflects the specialism of what Saunders some time ago called 'national Anthropology'. As he says: 'I would list an exceptional topical concentration on religion, symbolism and ideology; a strong emphasis on historical approaches ...' (see Saunders 1984: 449; see also Filippucci 1996 for an overall review).

The lack of focus and interest of many Anglo-American (or even Italian) anthropologists in kinship and gender studies sets Italy apart. As mentioned, there is no parallel with work carried out in Greece or Spain (see note 6) and on how kinship and gender, personhood, identity, biological and social ties are conceptualized, as explored by scholars such as Strathern (1992a, 1992b, 2005), Schneider (1980 [1968]), Wolfram (1987), or (with reference to the medical and new reproductive technologies) by scholars such as Ragoné (1994), Franklin (1997) and Edwards (2000). The Italian tradition is rather of a sociological-historical kind and is informed by a stronger interest in the family and its changing forms: see for instance the work of Saraceno and Naldini (2001),

---

7. As pointed out earlier, to date only a few anthropological studies with a focus on kinship (and gender) are located in industrialized areas (see Filippucci 1992, a Ph.D. study in Bassano, in the north of Italy).

Saraceno (2003) and Barbagli, Castiglioni and Zuanna (2003). This makes it difficult to align *Conceiving Kinship* with any similar Italian study: the multiplicity of research contexts, the comparative aspects and the diverse layers of investigations, address questions that current anthropological works on Italy do not seem to have addressed, either implicitly or explicitly. After the critique of the Mediterranean area-construct (see Box 1, 'Note on the Mediterranean Cultural Area Construct') very few anthropological studies have been produced opening up new lines of investigations, or adopting new and different approaches (see a rare example in Yanagisako's work 1991, 2002). On the contrary, just as many assumptions about the Mediterranean as appeared in the literature up to the late 1980s (despite the absence of much ethnographic evidence, as de Pina-Cabral 1989 has pointed out) continue to circulate today. Between the lines, they re-emerge in the notion of Southern Europe, now deployed by many scholars to re-address a distinct cultural area. Often, they are reinforced in a stereotypical manner in the context of what Gell calls (1999) 'seminar culture' (cited in Spencer 2000). The questions that we are left to ask, then, are: what cultural features characterize this part of the world? How can we avoid reproducing 'old-fashioned' and unsubstantiated assumptions? What is there to be explored beyond the confines of the North/South framework? What is there to be said of it, despite what is now of course an obsolete critique of the Mediterranean cultural area construct, in the absence of an alternative theoretical framework? In what ways is Italy (and by implication Southern Europe) also a different place from the one 'seen' by most Anglo-American anthropologists in their earlier studies as well as in more contemporary imaginings? Or, in other words, what kind of questions can or should we ask when we, as native anthropologists, but trained in the Anglo-American tradition, approach such locales? (See Llobera in Bestard-Camps 1991, who interestingly addresses the notion of 'Euro-Mediterranean'). Why is there a problem in cutting Europe in half or in quarters (North/South/East/West) when we begin exploring how kinship is conceptualized?

BOX 1

Note on the Mediterranean Cultural Area Construct

Overall, the anthropological approach to Italy up to the early 1990s, with its almost exclusive focus on peasant and small communities, underdeveloped areas, land and family structures (mostly located in the south), rather than on urbanized and developed areas, cities and more cosmopolitan settings and practices, has fuelled a perceived polarity between the 'Mediterranean' (which for some included both Christian and Muslim sides, see Gilmore 1982) as pre-modern and traditional, and 'Europe' as modern and advanced. Since the early 1950s, Mediterranean societies have been portrayed as bounded and unmodernized: 'half-way' between European gesellschaft-based (fully 'modern') societies and the kin-based societies of Africa (Herzfeld 1987) or '"half way" along the path of modernisation (Cole 1977: 358), combining nation-states, cities, and patchy industrialisation with large populations, strong labour movements, and powerful religions' (Filippucci 1996: 53). Mitchell summarizes well the emergence of an anthropology of the Mediterranean in the 1950s to 1960s:

> [It] represented a 'homecoming' (see Cole 1977) by scholars hitherto preoccupied by Africa and the Orient, the emergent sub-specialism of Mediterraneanism served to plug a discursive gap between Western self and non-Western other. Mediterraneanism, a term Herzfeld chooses to echo Said's Orientalism (1978) describes a corpus of empirical works and – more importantly – a body of theory aimed at describing the Mediterranean, but which in the process produces an ideologically-motivated discourse informing an anthropology that creates the rules for the management of the history – and social science – of the region, narrating the Mediterranean by itself, for itself (Herzfeld 1987: 65). Through this discourse, the Mediterranean is produced as *other* – not Europe, not even Southern Europe, but a category discrete from, and hence opposed to Europe (see also Cassano 2001). (2002: 4)

Within such discourse, and the construction of *otherness*, certain distinguishing cultural and social features or traits were isolated, so as to give rise to the notion of a distinct cultural area – in opposition to 'Europe'. In the 1960s Pitt-Rivers and Peristiany (1965, 1976; but see also Schneider 1971) proposed the distinctiveness of the Mediterranean, pointing out the 'honour and shame' syndrome as a marker of unity (see also Gilmore 1987). The Mediterranean was constructed, as Gilmore explains, as:

> ... A bundle of socio-cultural traits ... briefly, these are: a strong urban concentration; a corresponding disdain for the peasant way of life and for manual labour; sharp social, geographic, and economic stratification; political instability and a history of weak states; 'atomistic' community life; rigid sexual segregation; a tendency toward reliance on the smallest possible kinship units (nuclear families and shallow lineages); strong emphasis on shifting, ego-centred, non-corporate coalitions (Schneider 1971; Schneider and Schneider 1973, 1972, 1976); an honour and shame syndrome which defines both sexuality and personal reputation (Peristiany 1965); The dynamics of community life also bear many affinities. Most villagers share an intense parochialism or *campanilismo*, and inter-village rivalries are common. Communities are marked off by local cults of patron saints who are identified with the territorial unit. There is a general gregariousness and interdependence

BOX 1: continued

of daily life characteristic of small, densely populated neighbourhoods. Mediterranean communities also feature similar patterns of institutionalised hostile nicknaming (Barrett 1978; Brandes 1975; Cohen 1977; Iszaevich 1980). The evil eye belief is widespread (Appel 1976; Garrison et al. 1976; Herzfeld 1981; Moss et al. 1976) Religious and ritual factors, which one would think so dissimilar [as they unite Christian and Muslim!] have also been seen as providing correspondence. ... Marriage patterns seem to vary greatly at first view, but on closer inspection underlying commonalities again emerge. Hughes identifies the practice of dowry as a distinction that 'still sets the Mediterranean apart' (1978: 263). (1982: 179)

As much as notions of a pan-Mediterranean unity have subsequently been questioned and criticized (see Herzfeld 1987; Campbell and de Pina-Cabral 1992; Mitchell 2002), the construction of an anthropology of the Mediterranean has contributed to specific imaginings, with important theoretical and methodological repercussions – starting with the scarce attention to urban locales. Macdonald sums up the problem, drawing on earlier criticism:

A number of anthropologists attempted overviews of the social anthropological study of Europe ... They were particularly critical of anthropologists' concentration on small rural communities of Europe ... European anthropologists were accused of engaging in an 'analytically restricting village fetish' (Gilmore 1990: 3) and of ignoring in the process much of that which was characteristic of European life. (1993: 5)

Goddard also points out the implications of such an approach and, consequentially, the limits of generalization, when there is an absence of research in alternative settings:

The paucity of urban research has meant that generalisations concerning 'the Mediterranean' have been based on studies of small rural communities. This has meant that much of what has been claimed by anthropologists as typifying the area has little resonance in studies of urban situations. (1996: 8)

BOX 2

Note on Catholicism as a Framework

Part of the Mediterranean cultural area construct rests on the religious divide between the Catholic and the Protestant ethic. A common assumption is to take Catholicism (and Protestantism) as *tout court* distinctive social and cultural signifiers. The dichotomy is often taken for granted, and its implications little questioned (see Mitchell 2002). Catholicism, although the main religious denomination in Italy, requires further analysis due to Italian dissonance towards it: atheism, agnosticism, scepticism, belief but lack of practice, signify important differences (see Pratt 1996). Della Pergola, in the 1970s, made a distinction between 'traditional Catholicism', distinctive of the pre-conciliar time, and 'progressive Catholicism',

BOX 2: continued

distinctive of the Catholic culture of the Second Vatican Council (1975, cited in Allum 1990). Allum writes:

> Since the vast majority of Italians are Catholics, it might be thought that this 'official' Catholic culture is Italian culture and vice versa ... however, it is as well to bear in mind that, if, at the end of the war, some 95 per cent of Italians were baptised into the Catholic faith, the proportion of practising Catholics was significantly less (circa 60 per cent) ... Since the 1950s with the rapid growth of secularisation, practising Catholics have declined, becoming a clear, if strong, minority (circa 30 per cent), and the nominal and non-Catholic segment of the Italian population has grown correspondingly. (1990: 80)

In post-war Italy, Catholicism has transformed itself at several points in time, and in various ways each time in conjunction with important societal transformations – for instance the economic transformation that occurred between 1958 and 1963: the so-called 'Italian miracle'. The Italian miracle represented a sudden and accelerated economic growth that led Italy into the G8, and which generated major social change in society, including a radical reform of Catholicism. Again, Allum (see also Forgacs and Lumley 1996) illuminates the point fully:

> It was evident to the more prescient in the 1950s that the rapid economic and social change of the 'Economic Miracle', including the mass migrations of population from countryside to town and from agriculture to industry that this set in motion, *were increasingly undermining the credibility and appeal of traditional Catholic culture*, founded in blind obedience to the Church hierarchy and on the virtues of 'rural civilisation'. Mainly thanks to French Catholic ideas... small groups of Catholic intellectuals... had, since the 1930s begun to formulate *a more modern version of Catholic Culture*. (1990: 89, emphasis added)

Similarly, Catholicism in Italy needs to be understood in the political arena as post-war Catholic parties have been challenged by a complex set of political phenomena (see Lumley 1990; Lumley and Morris 1997) as much as by the form of political, parliamentary and governmental representation. As Shore suggests, '[in Italy] the major parties are, in effect, *cultural systems*' (1993: 32, emphasis added). In such a landscape, contrasting values coexist, continually recombining a diversity of views, including secular ones. Diversification of political representations is 'structural' to the Italian body politic, which is also formed on a plurality of secular values and beliefs. It is the principle of Italian secularism – in all its forms – to defeat Catholicism. Spotts and Wieser explain:

> [In Italy] there is a greater acceptance of political diversity than in any other Western state, yet Italy has been marred by one of the world's highest levels of social discord ... *The population is nominally 97 Catholic and the papacy has for centuries played a central role in the country's life, yet today society is so secularised that the Church has almost no political influence.* (1986: 1, emphasis added)

Further, the Catholic ethic, Catholic political parties and Catholic associations have historically been opposed by two strong political, noteworthy traditions. The first is the Italian Communist Party, which was

BOX 2: continued

founded in 1921 and was notoriously the largest in Europe. Spotts and Wieser explain its feature well: 'It is a party that has lived between two worlds – the ideological world of Marx, Lenin, and Stalin and the practical political world of Western parliamentarianism and pluralism' (1986: 41). The significance of the Communist Party (which has been transformed only more recently) rested in its capacity to be a secular force and an alternative to Catholic values. As Shore points out, it replaced in many respects the role played by the Church (see also Kertzer, 1980, 1996): 'In many communities it [communism] has come to rival, if not supplant, the traditional role of the Catholic Church' (1993: 28).

The second memorable tradition that disturbed the Catholic Church and the propagation of its credo has been the Italian feminist movement, since the late 1960s one of the widest movements in Europe at the grassroots level. Bono and Kemp illuminate its features (but see also Passerini 1994):

> [There are] two crucial characteristics of Italian 'feminism': first, the presence all over Italy of many groups ... the attempt to create nation-wide networks, and the pluralism of the groups, the conflict and dialogue between them; and second, the importance of politics for Italian feminism, reflecting the strongly political character of Italian society in general. (1991: 2)

In sum, Italy is nominally but not overwhelmingly Catholic (thanks also to high levels of immigration). Catholicism and the Catholic ethic needs continuous historical contextualisation, and cannot be taken *tout court* as the analytical framework. De Pina-Cabral makes a useful point in this respect, which is still left uncharted: '...What is expected of fieldworkers [is] specifically the need to consider historical evidence. The importance of social history is perhaps particularly clear to European anthropologists who are studying their own societies' (1992: x).

CHAPTER 2

RESEARCH IN PLACE: SHIFTING FIELDS OF ENQUIRY

Introduction

This chapter sets out the investigation. It describes the social relations that made fieldwork possible and the various field sites: clinicians operating in private clinics of assisted conception, infertile heterosexual couples suffering from imparted infertility and undergoing programmes of gamete donation, infertile couples undergoing programmes with their own gametes, couples with no vested interest in assisted conception, couples suffering from infertility and attending programmes with their own gametes and, finally, adoptive couples. The aim is to reconstruct the research setting, contextualize certain research choices and the methodological approach, including what I have called the 'without-method approach'. This is a strategic device that has made it possible not only to gain access to clinics of assisted conception at a time when, because of the legislative vacuum, clinicians were particularly wary and protective about their practices, but also to gain trust and to overcome their discomfort about giving access to a scholar working within a discipline which they knew little about. This chapter, by offering a full account of the field, is part of the ethnography itself.

BOX 3
Note on Social Anthropology in Italy

In Italy, social anthropology and the practice of prolonged fieldwork is not fully established as in British and American academia. Social anthropology is not followed as a course of studies in itself but is often (still) taught as an adjunct within other disciplinary courses under the name of 'anthropology' or 'cultural anthropology'. As Saunders points out, it often 'spills over' into closely related disciplines such as history and sociology (1984: 449). The

BOX 3: continued

Italian tradition has most of all a history in ethnology and folklore (see Saunders 1984; see also Filippucci, 1996). As a result, at the time of fieldwork, I did not encounter great familiarity with debates over the future of the discipline (see Ahmed and Shore 1995), with the relatively more recent trend of anthropology in familiar places, 'at home' (Strathern 1987; see also Jackson 1987; Okely 1996), and with emergent forms of theoretical interest and anthropological enquiry of the kind that one finds in the latest collections of works such as Ong and Collier (2005). The idea of an anthropology of science and medicine was also hardly known beyond distinct academic circles – outside such circles it was often perceived as a hybrid.[1] Nonetheless, the word 'anthropology' is often used in parlance by anyone – not just by academics and other intellectuals. One can frequently hear the expression 'in anthropological terms' or 'anthropologically speaking'. It is difficult to say what is meant by that: several meanings are attached to such expressions. The word 'anthropology' is mostly used because of its etymology, and not to indicate a disciplinary perspective and field of studies. Within this broad context, some informants (both couples and clinicians) seemed to have degrees of familiarity with the idea of anthropologists working with people living in remote undisclosed areas (but do they still exist?), or closer to home, studying folkloristic practices. Many more, possibly the majority, were genuinely surprised that social anthropologists work with live humans (and not bones, genetic material and/or animals), the association being in the best of cases with physical or biological anthropology; in other cases ethology or zoology. As a result, anthropology always needed to be negotiated – often by negation – what anthropology is not, rather than what it is – and by comparison – how anthropology differs from other disciplines and practices (ultimately, how it differs from other subjects informants are more familiar with and may have a higher degree in). To say that my aim was to explore conceptions of kinship (come le persone concetualizzano le relazioni della parentela) through the lens of assisted conception, and particularly programmes of gamete donation, could not be more meaningless to many of them. To claim that I intended to do so with a comparative study between heterosexual couples suffering from impaired infertility and lesbian and gay couples making families by donation was even more incomprehensible. No one in 'real life' thinks in those terms, or asks those kinds of questions. I was always asked to find better ways to explain what I would do with my interviews, with accounts of infertility and life histories. I doubt I was successful in most cases.

---

1. I have encountered poor familiarity with the discipline of social anthropology during my current fieldwork in the UK amongst media professionals, interest groups and science exhibition experts. Most informants have overall quite an unclear idea of the discipline, and often associate anthropology only with biological anthropology (see Bonaccorso, forthcoming)

Multiple Investigations, Sites, Informants

This is a multi-sited ethnography that engages in multiple investigations in the field of assisted conception in Italy (see Hastrup and Olwig 1996; Marcus 1998; see also Gupta and Ferguson 1997): it simultaneously explores the kinship narratives of heterosexual infertile couples attending programmes of gamete donation (egg and sperm donation) as well as those of lesbian and gay couples making families by donation with the aid of informal (non-clinical) networks. It also explores the views of a collateral group of informants: couples without a vested interest in assisted conception, infertile heterosexual couples attending programmes with their own gametes in public hospitals and, finally, heterosexual infertile couples, all of whom have unsuccessfully tried AIH (artificial insemination by husband) and/or IVF with their own gametes and, when left with no other alternative then to undergo a programme of gamete donation, have chosen adoption (see table 1). These further, collateral investigations, were not part of the initial fieldwork plan, but developed along the way partly due to new observations and unexpected circumstances that enriched the investigation and partly due to external factors which interfered with work in the clinics – indirectly favouring new explorations. On more than one occasion I was 'suspended' by clinics due to external events that made clinicians wary. The

Table 1 Sites of Investigation

	Informants	Treatment	Professionals	Legislation
Inside private clinics	Infertile heterosexual couples undergoing programmes of gamete donation	Programmes of gamete donation (AID and IVF with egg/sperm donor)	Private clinicians and medical staff	No statutory legislation
Outside private clinics	Infertile heterosexual couples undergoing programmes with their own gametes Lesbian and gay couples Adopting couples Couples without vested interest	Programmes with couples' own gametes (AIH and IVF)	Public clinicians and medical staff Patients' organizations Experts in assisted conception (academia, law etc.)	Ban on programmes of gamete donation in public hospitals

NAS – a special police body – started to randomly check clinics when cases of malpractice were reported by the media and this made all clinicians uncomfortable with having an external presence in the clinic. The significance of these further explorations will emerge throughout this work but, essentially, they highlight from different angles and perspectives certain problematic aspects of gamete donation programmes.[2] Additionally, the study is enriched by the perspectives offered by visits to institutions and organisations operating in the field of assisted conception and clinics located in various parts of the Italian peninsula – which tend to have a diverse clientele. There I interviewed head-clinicians and senior staff. I also visited and interviewed head-clinicians working in public hospitals, in centres for infertility treatment, which did not offer programmes of gamete donation but were allowed to offer treatment with couples' own gametes (in line with Degan's administrative act, 1985).[3] I also visited patients' organizations which aimed to inform infertile couples (at the time they were mostly supported by clinicians themselves, making them non-independent). Similarly I also interviewed several non-medical experts in the field of assisted conception such as academics, lawyers, feminists and media commentators with the aim of exploring the diverse sides of assisted conception, and particularly of gamete donation. Due to long-standing attempts to resolve the legislative vacuum, there was and still is (although for different reasons) a vivacious and robust political debate on the subject. In the same period I kept a close eye on the work of the Commission for Social Affairs (which at the time was working on a new text for legislation), on the Office of the Ministry of Health and the National Order of Clinicians to keep informed of any new developments for clinics, clinicians and the programmes themselves.

Main Investigation

An Anthropologist in the Field of Assisted Conception in Italy: Strategizing Access

On arrival in Italy, and Milan where I was based, the priority was to enter the field of assisted conception as an *anthropologist*. I soon realized that this was going to prove a difficult enterprise for two main reasons: first of all

---

2. I should warn the reader at the outset that these further investigations are partial compared to the main investigation of infertile heterosexual couples undergoing programmes of gamete donation and lesbian and gay couples making families by donation. In this work I have not dedicated a chapter to them – when possible the data are summarised in footnotes. Their relevance rests in the role that they play in informing my overall analysis and concluding remarks. They have provided the broader context of this ethnography, helping to reveal the kinship narrative that lies in the background of many couples' choices. For this reason they are extremely valuable.
3. A more complete account of the legislative state of the art of Italian assisted conception is introduced later on and in Box 7, 'Note on the Lack of Statutory Law on Assisted Conception'.

because, as mentioned at the outset, the disciplinary field of social anthropology and the practice of 'prolonged fieldwork', is not widely understood. The notion of an anthropology of science and medicine was often unfamiliar to several of my potential informants – certainly to the clinicians and medical staff I encountered at the beginning of fieldwork. Many found it difficult to make sense of my interests; it sounded peculiar, an oddity, almost an eccentric claim that I wished to investigate assisted conception, infertility, IVF treatment, egg and sperm donation, and wished to spend prolonged time in the clinic. Clinicians wondered what kind of anthropology could possibly be done in clinics, labs, consultation and waiting rooms. What was so fascinating about such places anyway? What was so singular about what they said and did – and about what their couples said and did? What needed to be observed, taken note of, scrutinised in such detail? And if it was so interesting, how come they did not realize it? Clinicians were overall troubled by a discipline they knew too little about – by the idea of *prolonged observation* and *immersion in a given context*, which I endlessly tried to explain, increasingly perceiving degrees of puzzlement. Certain levels of disciplinary incomprehensibility and distance surfaced in obvious ways, reminding me of C.P. Snow's long-standing argument (1998 [1959]) of the separation between the humanities and the sciences. They were obviously unsettled by the prospect of having an anthropologist around – someone who looked for 'things' that they perceived as obtuse and, indeed, could not see themselves. Secondly, and more generally, clinicians were clearly unsettled by the prospect of having an anthropologist at the very specific political and legislative moment in which I approached them and wished to carry out fieldwork. They perceived, saw and experienced everyone as a potential nuisance and not as enrichment or as source of intellectual exchange. The lack of legislation on assisted conception and hence the unlegislated status of programmes of gamete donation made them particularly defensive about their practices, procedures and management of programmes. The legislative vacuum in which they operated created enormous tension (as much as it offered extraordinary freedom to sell programmes in a highly profitable market). They were nervous about exposing their self-made rules and becoming transparent – they were aware that they were acting entirely outside the frameworks that their Europeans colleagues had to abide by (see for instance the regulatory framework set out by the HFEA in the UK at the time). This created great unease, as they wished to minimize any risk of disturbance.[4] As

---

4. As explained later on, the lack of legislation left the private sector free to dictate the rules of Italian assisted conception and particularly of gamete donation programmes which were widely perceived as most controversial. Besides the social/cultural implications that they raise, they require specific procedures and ad hoc legislation to be put in place: for instance in relation to the selection, recruitment and screening of donors, conservation of sperm and eggs, and informed consent procedures. At the time the Italian private

commented at the time, it all sounded like heavy 'village politics' (Strathern 1998; personal communication).

Because of these factors, 'access' required quite a bit of negotiation and manoeuvring. First of all, I needed to obtain permission, but I also needed to avoid being granted it on the wrong terms, and then being silently turned into an external and distant observer, obstructed in various ways all the way along. I needed to avoid – as Augé would put it – being nowhere at all times (1995). This was a likely prospect, since clinicians might just decide to grant access on a formal level, so as to be able to claim to be part of the study, but obstruct in practice the actual work. In other words, access might remain a pure formality.[5] Or, as I was told at the first meetings, it could be granted and then dropped any time depending on various factors, which might not be made fully explicit to me. In other words, as I was kept in a state of alertness from day one, with little sense that I could count on agreements and being able to work in the clinics on a regular basis, I needed to make sure that I lost no opportunity that came my way. From the beginning clinicians made the rules very clear. They said that 'to be granted access' meant only that I could walk into the clinic and be recognized as 'the anthropologist'. It did not mean that I could work directly with members of staff, or that I could approach couples directly in any way. They pointed out that they controlled patients (as they called them) and that I would always need their authorization and consent to approach them just as I would need their consent to approach anyone else in the clinic. I could not take any initiative, and no one else in the clinic could take it on my behalf: access had to be renewed each time, each day, with each couple. I also signed a confidentiality agreement which secured anonymity at all times, and which they locked in a safe. From the beginning, clinicians also pointed out that the clinic might be unable to provide what I needed. They seemed to be wanting to be sure they had ways to obstruct research requests with potential obstacles – if they so wished. To obviate any complications, I put in place two strategies. With the first I made sure, whilst in the clinic, that I followed the rules as far as I could, even though they often seemed irrational. As part of this strategy I quickly learned what I had to say and what I should not; what kind of information it was better to omit and what not to omit. It is important to note that ultimately the adherence to rules and the control of what was said did not limit the enquiry. Given the context, this was simply necessary for gaining continuous access. With the second strategy, I made sure to

---

    sector exercised hegemonic power, lacking transparency as far as procedures and management of programmes were concerned; therefore any outsider was perceived as a threat as they could expose malpractice.

5. This also needs to be understood in the broader context of Italian social/professional relations, whereby it is always 'good' to take part in something that in principle can confer prestige in some form (for instance, by claiming to be collaborating in a research project). Of course, a certain degree of familiarity with the cultural context made it possible for me to anticipate much of what could easily have happened (see Strathern 1987).

cause the least possible 'distress' to clinicians; in other words, to make the research project trouble-free. This second strategy – which I call 'without-apparent method' – required that I had no specific research (and methodological) requests. With it, I did not give anyone any reason to build up resistance or claim that my research requests could not be satisfied (suggesting, as a consequence, that access should be dropped). I claimed that I wished only to meet couples suffering from infertility and attending programmes of gamete donation – and that this was my only criterion for 'selection'. I explained that couples' age, origin, background, education, class, profession did not matter, nor did it matter whether the couple was at the beginning of a programme, halfway through it or had just conceived. I insisted that I was keen to meet any couple undergoing a programme and willing to 'chat' with me. I pointed out that I had no constraints on the clinical history of the couple – it was not relevant what brought them to choose gamete donation in the first instance, and how many years couples had been engaged in infertility treatment overall. I explained finally that I had no limits on how many couples I met, as that would be decided in time. Most of all, I believe, I worked hard to keep my research brief *always* as broad as possible (research questions and aim of the study). I was thus as aspecific and asystematic as I could be. I adopted the most laissez-faire approach possible, and built a very loose research profile, which made it difficult for clinicians to claim that they could not provide what I needed, research wise.

Paradoxically, given the circumstances, both strategies worked out well for a number of reasons. First of all, I managed to maintain access virtually throughout the fieldwork, apart from a few occasions in which, as I mentioned, I was asked not to attend the clinic due to external scrutiny. Secondly, as I was generally undemanding, I conveyed a sense of reassurance to clinicians. I noticed after a while that the more aspecific and asystematic I appeared, the more clinicians relaxed. Through time, such an approach lessened their anxieties and made them feel less exposed, although I am sure they must have heavily criticized my apparent lack of methodology as – formally – they were indeed always looking for 'protocols', 'questionnaires', 'control studies'. The loose research profile I built was, from their perspective, a sign that I would not be able to grasp those tricky areas of Italian assisted conception that they wished to cover up. They were operating with no regulatory frameworks, and believed that systematic enquiry would be required to grasp fallacies. Finally, the without-apparent method approach turned out to be an effective solution from another point of view. In time, it allowed me to collect valuable data by offering real access to couples. Whilst clinicians initially screened and 'passed on' selected couples – couples they claimed to be 'the most suitable' (i.e. a couple that succeeds in a programme and therefore reflects positively on the clinic and overall rates of success) – slowly they lost interest in that exercise, which must have required thought and effort in the role of gatekeeper. As soon as they gave up, as they had nothing on which to base their selection criteria (since I had put no limits), they began to 'pass on' couples with different backgrounds and clinical histories – including many with a history of unsuccessful treatment.[6] As it was, it was almost perfect: I gained an overall picture of *who* actually chooses and attends a clinic of infertility treatment, and keeps undergoing one treatment after another until a baby is born. It became evident that couples choosing gamete donation are not simply 'self-selected'; they have a salient narrative in mind which shapes itself rather powerfully and differently from that of the other couples who constitute my collateral investigation (see later). This work is set to explore it fully.

Infertile Heterosexual Couples and Private Clinics

Once in Milan (where I was mostly based), I started to contact clinics located in the city and outside; and increasingly in other cities. With some difficulty I arranged meetings with head clinicians, passing several times through the screening of well-trained secretaries. At the very beginning it was often the case that head clinicians were extremely busy – with a patient, in the surgery, in a meeting, in another clinic, at a conference, on leave, or often just on the phone. At that time, no one would call back if I left a message.[7] When I was given access to my first clinic I attended regularly and started to meet couples, depending on the treatment of the day. When I was given access to a second clinic I would share my time and my days between the two. When I was given access to further two clinics I tried to organize day-trips to meet couples in each different clinic, but I should say I did less work in the latter two. Organization was difficult because I was never told beforehand how many couples I could meet that day in a specific clinic until I was physically there; when I asked clinicians

---

6. It was at this point that I realized that the strategy was working. For instance, in one clinic I found I could increasingly move around instead of just waiting in the room that was allocated to me. At one point, a colleague from Cambridge came to visit me and I was able to freely introduce him to staff, and to walk around the clinic. Before, that would have been inconceivable. More importantly, I was allowed to sit in the laboratory with the biologist and generally wander about. Another clinic suddenly issued little brochures stating that it was collaborating on an international research project!

7. The annotation of lack of interest and availability is significant as the situation changed over time when by word of mouth head clinicians across various clinics came to know about my work. A year later the situation was substantially reversed: head clinicians made themselves fully available, and would call back if I left a message, whether I knew them personally or not. On different occasions, for instance in clinics in Turin and Palermo, I was offered immediate access without having to ask or negotiate. In Turin, the clinician suggested that I start immediately, that at the end of the (first) meeting I stay and interview the women waiting for treatment in the waiting room! He also offered me the opportunity to attend surgical interventions there and then. In Palermo, another clinician offered access any time: 'You can come here and do all the research you want. You will be my guest', he said.

if it would be possible to ask couples in advance if they were willing to meet up, clinicians made clear their unwillingness to do so. All they could do, they always pointed out, was in the course of a consultation 'say a word' and 'suggest' to couples that they meet me. I, of course, was denied direct access. Although, as mentioned, clinicians' attitude changed over time, there were areas that they never let go completely: direct access to couples was one of those. (To justify themselves, clinicians often pointed out that they knew how to convince couples – a very typical statement would be *'Lasci fare a me'* – let me do it). As a result, throughout fieldwork, I could not contact couples beforehand with the help of secretaries; I could not circulate a letter; I could not approach couples in the waiting room; I could not contact them by phone following a letter sent from the clinic. I could not follow the example of other scholars who have done work in similar clinical settings (see Ragoné 1994; Franklin 1997; Cussins/Thompson 1998, 1998a, 2005; Becker 2000). As I had decided to be undemanding, I could not put forward any requests but only play it by ear. My first contact with couples always passed through the clinicians who were the gatekeepers.

The first meeting always happened in the clinic (clinicians provided a room for my interviews), in between or at the end of a medical consultation, examination and/or AID or IVF with donated gametes. A meeting lasted between one and one and a half hours, and couples were introduced to me very quickly by the clinician on call. In almost all cases, I had fragmented information about couples' clinical history, as well as at which point of treatment they were. The usual briefing would be 'this is an egg' (this is the case of an egg donation) or 'this is an *eterologa*' (this is the case of an artificial insemination by donor). Even when I was present at consultations, examinations and/or surgery, I had to work out everything by myself from the conversation as there was no way I would be provided with information beforehand. I would, though, be asked to wear a white coat: appearance of course did matter. Paradoxically, but interestingly, this complete lack of information turned out to be extremely advantageous. My first meetings were never influenced by clinicians' views on the case. If I needed to (and I often did), I would always manage to ask clinicians and medical staff for explanations or details later on, and they would usually offer interesting portraits on this or that case. No one ever missed the opportunity to make funny and crude comments on the couple and/or on the medical case. It is often the case that couples one meets in a clinic for infertility treatment and who attend programmes of gamete donation have been 'patients' for a while. The long-term nature of treatment gives clinicians ample 'material', so to speak, to base their views upon. Of course, as in any profession people can became quite cynical. This, naturally, was all quite fascinating: what was intriguing for me was very ordinary for them, and the reverse.

I have interviewed forty-eight couples, often wife and husband together, but also wives only, and more rarely husbands alone. This number reflects full, in-depth, completed interviews. However, not all interviews are equally informative and rich. Many are repetitive and predictable – their value rests in reconfirming the overall data. In-depth interviews were semi-structured – I chose about twenty themes to work through with couples. I always tried to hold conversations on the same themes with all couples, but of course couples were interested in different issues and, often, they decided the way they wanted to go. I redirected conversations only if they focused excessively on irrelevant medical details or exclusively on certain aspects of gamete donation. At the end of each interview I always annotated what seemed to me important aspects: the choice of specific words, intonation and emphasis; the way partners related to each other; the way they shared answers (e.g. 'individual' and 'common' answers); and, importantly, the way they related to me. In most cases interviews took the form of personal and intimate accounts. Couples were overall quite relaxed although there were circumstances in which one of the partners became emotional, cried or articulated profound sadness. Often, I noticed more openness in the first encounter than in subsequent interviews, though the latter was entirely a matter of choice (see Box 4). Thinking back about the sensitivity of the

BOX 4
The Second Meeting and Beyond

Although I knew clinicians did not want me to pursue contacts, I always gave couples my contact details and/or business card at the beginning of our first interview. It was unreasonable for me to ask couples to entertain such sensitive conversations without introducing myself formally, leaving a trace and offering the possibility to pursue the relationship if they so wished. I thought it would be a matter of courtesy and sensitivity, besides any professional interest. It remained somehow understood that it would have to be up to couples to contact me again, and that I would be delighted if they did so. Many couples did contact me, and it worked well. I am sure couples perceived that this was a tense ground with clinicians, and that it would have been difficult for me to take the initiative, despite the fact that I was asking for their contact details at the beginning of the interview. We always carefully avoided mentioning clinicians' position in relation to this – I suppose we all felt that there were some pending constrictions. I was very pleased, therefore, when couples contacted me again – this freed up the relationship and gave me an opportunity to reciprocate their willingness to share their experience of infertility and treatment in the first instance. It also gave me some confidence that they also gained something from our first meeting. Usually, the second meeting occurred between one and two months after the first interview: in a restaurant, in a bar for a drink, in a park. Only rarely I did meet couples in their own homes, despite my willingness to travel to where they lived. I pursued most of my contacts, only lessening contact when couples became particularly insistent suggesting that I could help them more than anyone else, confusing my role. With some couples I kept some form of contact throughout my fieldwork; a few couples updated me on various events: a loss, a new treatment, a new attempt, and in a few cases a scan and an examination. With some couples, whom I consider my main informants, I engaged in more continuous relationships.

conversations that took place, a part of me is still surprised at the depth and openness with which couples talked to me as an absolute stranger. But possibly this occurred precisely because I was an absolute stranger, not unlike an encounter on a train. It is not unusual to share very personal life histories with unknown listeners. With others one can be honest and direct as there are no implications. Couples could count on the fact that they would never see me again if they so wished. They could confine our encounter in time and space. It was entirely up to them to decide – to take or leave the relationship.

Comparative Investigation

Lesbian and Gay Couples and Informal Networks

The comparative investigation with lesbian and gay couples making families by donation was carried out outside clinical contexts, and outside the immediately visible lesbian and gay 'community'.[8] At the time of fieldwork, and still at the time of writing, lesbian and gay couples were not supposed to attend clinics for infertility treatment as clinics were not allowed to offer them their services. What used to happen in practice is a different matter altogether, but here I will not make any reference to couples I met through clinics or to couples who claimed to have attended clinics.[9] I will refer only to couples I met outside clinical contexts. I had previously carried out research amongst lesbian and gay couples (see Bonaccorso 1994) and had kept in contact with many of them. At the time

---

8. The notion of 'community' should not be understood as that of constituted entities of the kind described in ethnographies such as Green's (1997) or of the kind that one would find in the United States, particularly at the time (the situation has been changing rapidly in the last few years). This is particularly so when lesbian and gay couples making families by donation are under scrutiny. As will emerge in this work, lesbian and gay families by donation were at the time, and mostly still are, part of an informal and fragmented network.

9. Although the field of assisted conception was totally unlegislated, with the exclusion of the 1985 Degan's regulation, the National Order of Physicians in 1994 agreed to self-regulation, recommending that private clinics did not offer treatment to lesbian and single women. In 1994 a well-known clinician, who provided treatment to a lesbian couple (the case was widely reported in the national press), was expelled by the Order. However, in everyday practice, things were different. Some clinicians did offer treatment to lesbian and single women. I was made aware of it in my previous research, carried out between 1991 and 1993. I met and interviewed lesbian and single women who were 'secretly' attending clinics of assisted conception. They would visit the clinic with a male partner to avoid raising suspicion amongst general members of staff. In agreement with the clinician a false case of medical infertility would be construed. In this second wave of research such practices continued.

I interviewed couples who had had children from previous heterosexual relationships and then moved into a same-sex partnership; couples who had used artificial insemination by (unknown) donors; and a minority of same-sex couples who in agreement with opposite-sex partners had planned a child. I also had an independent circle of lesbian and gay friends, from when I used to live in Milan, on which I could count for help. Almost immediately on my arrival, and concurrently with getting access to private clinics of assisted conception, I contacted everyone I knew to establish contacts and regain *access* (I had been away since 1995). I put the word around that I was interested in meeting, if possible, new couples planning to have children.[10] I wanted to gain access to the highly fragmented and dispersed network of couples planning families by donation, or wishing to do so. In addition, a few months later, I put out a series of advertisements introducing myself, the work, and asking to be contacted in *Babilonia* (a national monthly magazine of lesbian gay culture), with the help of a journalist working there. Both approaches turned out to be successful. I began to receive phone calls from all over Italy (more than expected and more than I experienced in my previous research). These were from couples or singles either wishing to have a child, or planning to have one; some would phone as a go-between to put me in touch with a friend or someone else they knew about. After an initial chat on the phone, I would ask where I could contact the couple or the individual again, and under what restrictions. There were always restrictions and procedures that I had to follow, but this was no different from my previous research. All couples were afraid of exposing themselves publicly, for all sorts of personal, professional and also legal reasons. As always, I ensured secrecy and anonymity at all times. If the person agreed (and often they did not), I would call back and conduct an interview over the phone, or arrange to meet face to face. The latter was, of course, the preferred course of action for me, and when possible I suggested a meeting in Milan, or outside Milan, or in places such as Turin, Venice, Florence, Rome and Palermo where I was also visiting clinics. Similarly, I came into contact with more couples by writing articles or being interviewed on gay and lesbian matters by the Italian national press (e.g., *Io Donna – Corriere della Sera, D/La Repubblica delle Donne – La Repubblica*); or by talking on radio or television (e.g., *Radio Popolare; Radio Capital, Rai Due, Rete Quattro*).[11] Such *visibility* was

---

10. I could rely on the 'word of mouth' strategy to a certain extent as I had already carried out research in the early 1990s and, besides my contacts, I am known for having published the first academic work of lesbian and gay families in Italy (1994). At the time, the national press reacted immediately to its publication, associating it with the case of the tennis player Martina Navatrilova and her plans to conceive by artificial insemination, with the result that I become very visible. The text is still a point of reference as there is very little published on the subject.
11. In Italy, I am also a professional journalist, member of the National Order of Professional Journalists (Rome). This facilitated media appearances in the press and broadcasting.

extremely helpful: first of all, because in this way I was able to reach a much larger audience; secondly, because I was also able to pass on the message that I would maintain secrecy and anonymity, and those who contacted me could see how I treated the information I was given by other couples. It was self-evident that I was operating openly but, at the same time, when speaking, I never made reference to specific cases. As this was a primary concern for many couples, such media appearances did help substantially. For many lesbian and gay couples with children, and those in the process of planning families by donation, the main fear (in contacting and being contacted) is of being suddenly exposed, betrayed, and used as open cases by the media. A reassuring visibility can therefore be helpful. Finally, the very fact that I appeared in the national media conferred a sense of reassurance. Even though the media are often disliked, they are also seen as lending a certain status of seriousness to an issue.

I have altogether seventy-four records, including in-depth interviews, phone interviews and general contacts. Of the records, twenty-six are in-depth interviews (many consisting of more than one conversation), thirteen are one-off interviews carried out over the phone, nineteen are more general phone conversations, and sixteen are more general contacts consisting of quick phone calls with couples or individuals willing to say a few words but not happy either to be called back or to spend a long time on the phone. (I include in the total records those I talked to over the phone for a short period of time because 'numbers' indicate the extent of the response). Face to face interviews always lasted at least two hours, sometimes more. Interviews over the phone never exceeded one hour. Interestingly, in some cases the latter were more straightforward, as if there was less need to negotiate the 'personal' and it was possible to get to the crux of the matter in more immediate ways. In both face-to-face and phone interviews I worked, as with the heterosexual infertile couples, on a number of themes. The aim was to collect comparative data. However, conversations with lesbian and gay couples (more than with other informants) always incorporated the wider context in which lesbian and gay couples live and make their choice of making families by donation. Their accounts therefore in a sense reflected more the very specific point in time at which they took the decision to initiate such projects: these are complex enterprises that require the participation of others with whom couples enter in unusual negotiations, outside of taken-for-granted practices of conception, reproduction and kin relations, with potentially all sorts of implications and complications for all those involved and, primarily, for the children to be born.

Collateral Investigation

Whilst in the field the decision to carry out additional investigations materialized. I carried out fieldwork amongst couples without vested interest in assisted conception, infertile heterosexual couples attending public hospitals and undergoing programmes with their own gametes, and infertile heterosexual couples who instead decided to adopt. Whilst the first of these three collateral investigations, amongst lay couples, was entirely planned, the latter two (amongst couples attending public hospitals and adopting couples) arose by chance.

Couples without Vested Interest

Whilst I was carrying out fieldwork I became interested in the comments that so-called 'lay people' would make on cases of assisted conception. These were people whom I encountered in all kinds of social contexts. It was quite striking the general, almost overwhelming, interest that the topic of assisted conception provoked – how far for instance it could catalyse attention and monopolize discussion at a dinner party – but also the kind of knowledge that many people seemed to have. The work of Edwards (1999 [1993]; 2000 retrospectively),[12] in the north of England, of course, lies in the background of these reflections. I soon began to wonder if I should not pay more substantial attention to such comments, and became intrigued by the prospect of investigating more formally what 'people out there' (outside clinical contexts and without known problems of impaired infertility) think about assisted conception and the practice of gamete donation in particular. I therefore decided to turn general – dinner party – conversations into *data* and address questions of the kind: 'What do "people out there" think about using unknown eggs and sperm to make babies?', 'What do they think about replacing their own genetic material with that of third parties?', 'What are the broad cultural and social constructs that they draw upon to sustain or reject such choices to conceive in the face of infertility?' With these broad questions in mind, I organized a formal investigation with couples without vested interest. I organized it through the Trade Union of Milan (*Camera del Lavoro*), contacting a senior union leader. The Trade Union was the perfect place, as was the senior leader, who in the words of a friend, 'knows half of Milan'. I gave him some indications: my aim was to 'match' as much as possible the range of backgrounds found in clinics of assisted conception, especially in terms of education and profession. These two factors are quite determinant in the construction of Italian social identity, more so than social class per se (see Ginsborg 2003). I hoped to be able to collect the views of people with a similar background to that of the couples met in the clinics. Only a few days later, to my amazement (given my experience in clinic and the time that it took to organize each meeting), I received a well-typed list of names and telephone numbers of people I could freely contact. To the list I added personal contacts, friends of friends and ex-colleagues (working in the national press and at the university). I also included the caretaker of the place where I lived, the baker down in the street, the hairdresser and the newsagent – in the clinic I had met couples with similar backgrounds.

---

12. Of course, reference to the latest work of Edwards (2000) is made a posteriori as at the time of fieldwork it was not yet published.

Altogether I conducted forty interviews, of which twenty-five constitute valuable data (seven are poor and eight incomplete). Interviews were quasi-structured – the form was that of a conversation on agreed topics: I would insistently and openly ask for views on specific issues. These were all people with whom, explicitly and deliberately, I could address matters I was concerned with. Interviews mostly lasted one hour, although to my surprise sometimes they could last up to two hours. Overall, people seemed very keen to enter such conversations and many had a lot to say on most specific cases and/or scenarios I would pose. Investigations into hypothetical situations are always somehow immaterial; yet, I think, they illustrate how people talk about controversial issues as much as the actual experience and knowledge they draw upon to articulate their views.

Heterosexual Infertile Couples and Public Hospitals

As mentioned above, the topic of assisted conception always triggered the attention of people in various social contexts. Similarly, it also offered an opportunity for people to report about cases of relatives, friends and acquaintances who had encountered the obstacles of infertility and had difficulties in conceiving. On all those occasions I, of course, always listened to the account and took the chance to ask if it would be possible to meet the person (or the couple) in question. Surprisingly, it often turned out that it was! These were always cases of infertile heterosexual couples attending public hospital and undergoing treatment with their own gametes. I never met anyone who would report a case of someone else undergoing treatment with donated gametes (this, as will be explained later on, turned out to be a salient data in itself). In this way I managed to contact thirteen couples attending public hospitals with centres for infertility treatment, undergoing either AIH (artificial insemination by husband) or IVF with their own gametes, very often at the beginning of treatment.[13] Although, of course, this was not like working in a public hospital (where I could obtain access only if I shared my data with *specializzandi* – postgraduate medical students – as I was told in several occasions during meetings with head-clinicians running public infertility clinics), these encounters were illuminating in many respects. These couples offered the opportunity to explore what occurs before couples reach private clinics of assisted conception. With them, I was able to trace the path that brings couples to private provision of treatment, but especially to the choice of gamete donation. I compared views on the choice of treatment with own and with donated gametes as well as

---

13. The very fact that these couples attended public hospitals implies that they would not be undergoing programmes of gamete donation, as these could not be provided there. As already mentioned, this was the only legislative measure in place at the time: the public sector was forbidden to provide treatment with donated gametes following Minister Degan's administrative act (1985). The private sector, which was totally unregulated, could – conversely – offer any treatment. This will be fully explored in Chapter 5.

experiences of treatment in the public sector, on the circulation of information about infertility, treatment and various programmes (including those involving the provision of gamete donation services offered in the private sector), and on clinicians' conflict of interest between the public and the private (many clinicians worked in public hospitals in the morning and in private clinics in the afternoon, shifting couples around).[14] This was all very interesting: it was the reverse of the sort of accounts that I collected in private clinics. All couples I met in the private sector had experienced the public sector at some earlier stage, and had followed a similar route: they all attended public hospitals, they all tried AIH and/or IVF with their own gametes and, when unsuccessful, decided to undergo programmes of gamete donation. Importantly, these encounters brought to light how infertile heterosexual couples treat infertility and what they conceal about it. It soon became clear that couples I would become aware of by word of mouth were only those who were undergoing treatment with their own gametes. That's precisely why they would be happy to *openly* share their experience with others outside clinical contexts. They had nothing to hide as they had not yet landed in the world of gamete donation.

Infertile Heterosexual Couples Who Adopt

In my last period of fieldwork the possibility of meeting infertile heterosexual couples who adopted or were in the process of adopting arose completely by chance, very much in the same way as my contacts with heterosexual couples attending public hospitals. I was talking to a friend about my work in private clinics and she described a friend who had just adopted, after failing several times IVF with her own gametes. The friend and her husband were resolute that they would never enter a programme of gamete donation. Of course, this was fascinating as it was exactly a reverse account of the kind I collected in private clinics. By now it had emerged vividly from my work with infertile couples choosing gamete donation that they did not want to adopt, and that adoption would be a very last option. For the majority of couples undergoing programmes of gamete donation, it was not an option at all. I, of course, asked if I could meet the friend in question and, as in the case of infertile couples attending public hospitals, the meeting was welcomed and we met. Soon after, I was informally introduced to other couples who were part of a self-help support group for adoptive parents in the outskirts of Milan.

---

14. At the time of fieldwork the Minister of Health, Rosi Bindi, worked to introduce a new system whereby, as in other European countries, clinicians had to choose whether to practice in the public or in the private sector. Later on, Minister Bindi managed to successfully put in place a fair system; however, it was soon dismantled again by the Berlusconi government, creating a new wave of malpractice and conflict of interest.

In all I interviewed ten couples who had chosen adoption after AID and/or IVF treatment with their own gametes failed. Six couples had just become adoptive parents at the time of the interview, four had received *l'idoneità* (clearance) and were expecting to be given a child shortly – they all hoped for a newborn or young child. Once again, these encounters turned out to be fascinating, although the number of couples was limited and the investigation partial. These couples, like couples attending programmes with their own gametes in public hospitals (most of whom were resolute in refusing gamete donation), illuminated much of the main investigation and the data collected with couples in private clinics. With them I was able to investigate assisted conception from the perspective of those who drop out of treatment. More saliently, I was able to explore issues of biological inheritance and genetic make-up as well as 'What is a relative?' from an altogether different angle. These couples offered me a bigger picture of why it is possible for some, like them, to make a different choice altogether: to a priori exclude gamete donation and go for adoption.

BOX 5
Note on the Use of Ethnographic Data

The ethnography (main and comparative investigation) presented in the following chapters makes ample use of quotes precisely to reproduce the intensity of certain cultural notions that couples attending private clinics and undergoing programmes of gamete donation, and lesbian and gay couples making families by donation, use widely. Without such extensive use of quotes, and just by reporting such repetitiveness and predictability, it would be not possible to render the powerful work of culture and how it is reproduced. This is what gives strength to the ethnography. Throughout the text, quotes from twenty-nine heterosexual infertile couples and twenty-five lesbian and gay couples are used (a brief profile of couples can be found in Appendix II and Appendix IIa).

Reference to collateral data will be made in footnotes, whenever necessary, so as to elucidate some aspects of the main investigation and offer additional material for reflection. More extensive reference to this material will be made in the concluding chapter.

BOX 6
Note on the Ethnographic Data

The kinship narratives presented in this work have a particular relevance because they have been collected at a very crucial time in the history of assisted conception in Italy – a time of legislative vacuum. The lack of constraints on which practices of gamete donation took place makes such narratives especially revealing as both couples and clinicians, although differently placed with regard to assisted conception, were left to operate within self-made and self-serving frameworks. Narratives are not obscured

BOX 6: *continued*

or channelled by the official statutory framework that exists in other parts of the world where assisted conception is practised.

As well as being revealing, the material is sensitive. It has not been published in full before now (see Bonaccorso 2004a; 2004b) because it seemed unmediated and too close in time to the real experiences of couples and, by implication, clinicians. It hints at practices that, at the time, and particularly with the subsequent regulation of private clinics, amount to malpractice. Some of the clinics I visited and worked with have been going through difficult periods and now practise under different names.

The material is also sensitive for the couples who feature in this work and I have omitted potentially hurtful material from interviews; for examples with donors who reveal their motivations, or with clinicians on their practice of donor selection, screening, payment and number of donations per donor.

As well as being careful with the choice of which data to present, I have also, of course, followed the anthropological tradition of concealing the identity of *all* informants.

CHAPTER 3

HETEROSEXUAL COUPLES: LIFE PLANS, IRREVERSIBLE INFERTILITY AND THE CHOICE OF A PROGRAMME OF GAMETE DONATION

A Case: Anna and Artificial Insemination by Donor

Anna (married to Paolo)[1] is thirty-three years old. She works as a manager in a fashion company in Milan. Her husband is infertile. I meet her for the first time while she is lying down on the gynaecologist's bed. In the room there are two clinicians. I am wearing, as they are, a *camice* (white gown). Before entering the room I was asked to wear it by one of the clinicians: 'you need to look like one of us', he explained. When I go in, I walk towards Anna, introduce myself and smile. She smiles too. I then stand up close to the bed, and do not feel like talking. I am new to the place as much as to the medical procedure. I feel emotional; I have never met Anna before and I find it quite extreme to be standing close to her in such an intimate moment. After a while, when I am still silent, she asks for my hand. I am not sure Anna has made a distinction between me and the clinician; I am not so sure she has captured the few words I said when I introduced myself: 'I am a social anthropologist. As Doctor X explained I am here for a study on couples who find it difficult to have babies.' The atmosphere is thick, the room is dark: there is only a violent purple light that illuminates the gynaecological bed and a little lamp placed on the clinicians' desk. Anna's body seems rigid and nervous, the face and the legs are contracted. She holds my hand so tight that it is almost painful. She is undergoing AID (artificial insemination by donor). She has already undergone three insemination cycles[2] without becoming pregnant. She is

---

1. From now on, every time I refer to a couple I will put the name of the partner in brackets. For a brief description of couples, see Appendix II.
2. Each insemination cycle consists of two or three inseminations; the number varies depending on the clinic.

classified as a regular. As soon as the clinician leaves the room (he moves to the lab to select the frozen donated sperm) Anna starts talking. She talks with a weak voice, as when one speaks to oneself. She says:

> I hate to do this. It is unbelievable what one has to do to have a baby [*ha dell'incredibile quello che uno arriva a fare per avere un bambino*]. If I think that there are so many people killing their children [she refers to elective abortion] this is unbearable and do you know where my husband is? At a meeting. Would you believe that?

Since I met Anna, she has undergone several AID due to her husband's infertility but contrary to expectations she has not become pregnant. Recently she has been given a diagnosis following that of her husband's: 'unexplained infertility'.[3] Anna's predicament, emotional pain and suffering is representative of what is felt by many other women and couples. She repeatedly reproduces the same cycle: new treatment, failure, and the same hope for a pregnancy, a baby and a family. However, her words also indicate some ambivalence towards the overall procedure and seem to express discomfort particularly with the idea of a donor's aid.

Introduction

This chapter highlights the time of dreams and expectations prior to discovering infertility and the time of compromises once irreversible infertility is diagnosed; a time when couples have to think about alternative ways to make their own baby and start a family. At this point couples not only have to change perspective and life plans – but they have to come to terms, slowly and painfully, with what that entails. They have to re-write their own life history, revisit assumptions and expectations that were previously taken for granted. They have to re-conceptualize their own procreative and family narrative and adjust to what they call *la realtà*: 'the reality'. To do so, they go through stages. None of them arrive at a programme of gamete donation from nowhere. There is a progression of thoughts and actions. Along the way, in between the discovery of infertility, its acceptance, the search for solutions and the actual programme of gamete donation, couples seem to lose parts of themselves. They experience loss and disillusion, a sense of failure and a profound sadness for what could have been, but is not. It often takes years before couples tune in to their new life circumstances. The decision to opt for gamete donation has wide implications. To put it crudely, ordinary couples choose most extraordinary ways to create the most ordinary families; as will become clear, these particular couples, who choose

---

3. It turned out later that Anna had been diagnosed with unexplained infertility prematurely. Cases of unexplained infertility affect approximately between 10 to 15 per cent of couples (American Society for Reproductive Medicine 2005).

programmes of gamete donation, are looking for a *life like everyone else*. This initiates a complex process.

Planning Our Life, Planning Our Children

In an Italian clinic for infertility treatment one meets very diverse couples, from heterogeneous backgrounds, class, education, wealth and political orientation.[4] Overall, couples appear to portray the general social and economic variation found between social classes in any urban Italian milieu (see Ginsborg 2003 for an overview on Italian social classes).[5] However, two features characterize the Italian scene: couples are mostly of Catholic denomination [6] and of Italian nationality. Non-Italian nationals who attend clinics of assisted conception are often (wealthy) couples who would rather follow a programme abroad to minimize visibility in their own country (see Nielsen 1996 for insights on reproductive tourism). Despite the diversity of backgrounds, these couples seem to have lot 'in common'. Their life histories portray a progression of similar stages and life events. They all seem to have left very little to happen by accident.

There are certain conventions that they generally follow: at first they spend some years together or they are 'engaged', as they say. When they decide to marry (or in some cases cohabit, that is live together as if married) they have usually finished studying, have rented or bought a house, and have secured a job. They have made sure that the finances are solid enough to provide all that is necessary for the family and the children to come. There are great expectations about the level of achievements that need to be accomplished before having children, and particularly concerning the economic power that needs to be secured. It is

---

4. From now onwards I refer to heterosexual couples who attend programmes of gamete donation and that, as it will emerge, have a specific narrative in mind. I am not assuming that all heterosexual couples live the same sort of life and have the same expectations. Investigations amongst couples who refuse gamete donation and couples who adopt have brought to light other ways to envisage heterosexual partnerships.
5. The social and economic heterogeneity of couples found in private clinics of assisted conception seems to be an Italian feature. Ragoné (1994), for instance, points out that American couples attending surrogacy programmes are middle to upper class. Davis-Floyd and Dumit claim that: 'who has access to these technologies and who does not is often dependent on who has money and who does not' (1998: 7).
6. In Italy the main denomination is Catholicism, which was the state religion until 1984. In 1984 a revision of the Concordat formalized the principle of a secular state. However, importantly, denomination does not imply belief and faith. Most couples in this study define themselves as generally Catholics, but non-practising, non-believers, atheist or agnostic. A minority classify themselves as practising Catholics.

because of long-term planning that couples come to a decision to have a child when in their thirties and increasingly more towards their forties. They often discover that they have difficulties in conceiving after many years of being together and come to assisted conception at a late point in their relationship.[7]

> Marta (married to Massimo) explains how it works:[8]
>
> We married whilst I was finishing university; my husband already had a job. We waited a few years before having a baby. I wanted to finish my studies and find a job as well and we wanted to buy a house. I was terrified of having children at that point. We also wanted time off for ourselves.

Other couples report a similar experience.[9] Irene (married to Alessandro):

> I wanted very much to get married, I already had my own flat. I was living there by myself. Alessandro was still living with his parents and moved to my place only when we married. Before he would stay for the week end, when he got a good job, we married. We could have waited longer. We did not want to have children at that point, we liked the idea of being the two of us for a few years.[10]

This is how it ought to be done. Couples take their time before settling down and having children, but they all seem to assume they will have children. The question is when, not whether a child is the ultimate desire; a child is the most beautiful life event and everyone expects it. As these couples find it difficult to think anything to the contrary, the desire in itself is never doubted, even less (what they consider) the naturalness of the event. Couples claim that to have children, to conceive them, and to be a parent is *natural and normal*. As Becker points out for the American case it is the achievement of the cultural ideology of 'normalcy' that 'encompasses all forms of parenting, but there is an ideology about biological parenthood because that is so closely tied to American notions of what is normal' (2000: 34).

---

7. This reflects a much more widespread trend in Italian couples who have children increasingly late (see Saraceno and Naldini 2001; Saraceno 2003).
8. Here, as in the chapters to follow, I will use quotes extensively. On the one hand, I wish to render the intensity of the feelings and emotions that the experience of infertility triggers and the constructs couples draw upon in conceptualizing their choice of gamete donation. On the other hand, I wish to recreate the same sense of repetition that I have experienced during my interviews. Altogether the material will reveal the power of certain cultural and social idioms as they emerge from the sameness of these narratives. I should remind the reader that a much better translation has been made from Italian since they appeared in my Ph.D. (2001).
9. Although I freely speak of couples these are *specific* couples who attend private clinics of assisted conception and choose to undergo programmes of gamete donation. They are thus self-selected.
10. This view is shared by couples who attend programmes with their own gametes, by those who adopt and by couples without vested interest too.

Sofia (married to Filippo) explains why this is the case for her too:

> It is difficult to explain something so *natural*. A woman always wants children. It is a maternal instinct. Oh, yes there are women that do not have that instinct but there are very few of them and I cannot really understand. The desire for motherhood [*il desiderio di maternità*] is one of those few things that one *has* [*possiede*] from birth. Without children a woman is not completely a woman. I feel incomplete [*incompleta*] as a woman and in relation to my husband and everybody else really. (emphasis added)

Giancarlo (married to Barbara):

> I feel that it comes from deep inside, like that I really need to resolve this part of my life now. I need to have my own children with my wife. *It is natural at this point*, in our relationship, after so many years of being together. I think *it would be very abnormal* if we did not want to have children together as a couple at this point. (emphasis added)

This holds for all couples. All couples make very similar statements and when asked if they feel any social pressure to have children they tend to respond negatively. They mostly say that to have children not only satisfies a physical as much as a psychological need, but also that it is a desire and ultimately *a choice*.

Irene (married to Alessandro):

> It is not something that people tell you to do. My mother never asked for grandchildren. It is me that wants a child. It is my body, and my head. It is physical and psychological. And it is normal to feel this way. Nothing to do with social pressures.

Gioia (married to Leopoldo):

> It is quite the opposite. There is almost a pressure not to have a child! Everyone tells you that it is expensive, and to wait and that you shouldn't have more than one. My father always says 'no more than one!' *but I know that he is dying to have a nipotino* [a grandchild]. (emphasis added)

Ottavio (married to Nadina):

> It is a choice – no doubt. It is not about the family or anything like that. I think our life would be quite empty without a child, you see. I would feel that I haven't achieved very much – that I have not done the right thing. I do not want to miss the few things in life that really count. I suppose many couples feel the same – don't they?

Discovering Irreversible Infertility

The shock of infertility is expressed in terms of 'why has this happened to me?' The feelings are of loss and disillusion: loss for discovering what one always takes for granted: a fertile body, able to reproduce others like oneself; disillusion at realizing, in a moment, that one is living a different life from the one he/she had imagined for oneself and one's partner. If at a very abstract level couples have envisioned difficulties, and may sporadically have reflected on the possibility that conception of one's own baby might not be immediate, no one has ever reported to have contemplated infertility as such. The prospect of infertility is just too painful to be even thought of. That's why, when it comes, it is shocking. Infertility is truly devastating news, and couples find themselves totally unprepared to deal with it.

> Costanza (married to Antonio):
> Somehow I assumed that I would get pregnant immediately, never thought about the possibility of not getting pregnant, absolutely never thought about not having my own children.

Infertility is so devastating that couples at times find it difficult to believe the results of the tests and the diagnosis, as in the case of Maddalena (married to Pietro):

> When I was told, I did not believe it, I kept saying to myself 'it must be a mistake'. I walked back into the doctor's room and asked to have the tests done again. The doctor started speaking, and went on for so long, but I could not hear him. I wanted to do the tests again.

> Her husband, Pietro:
> Infertility is the last thing you think of. It is 'one way', there are no remedies, it is impossible to accept it. It does not make sense. Why should people be infertile? It is unnatural. We are all born to procreate. I cannot come to terms with it.

The very moment of discovery, when one is given the news, is extremely painful. Life almost stops, emotions almost freeze. As Luca (married to Mariangela) explains, he could not breathe, he could not think, he could not work, he could not tell his wife:

> It just did not make sense to me. It is such a loss. It really breaks your heart. I stopped breathing for a while. My brain went off. I went back home and could not talk to my wife. She was expecting the results the following day. The doctor is a friend of us. He called me. I could not keep working. I had a meeting, but I could not stay. I left the office and walked around, it was December, 13th, '94. People were looking for presents. I found it disgusting, Christmas and all the rest of it. I was trying to figure out how to tell it to my wife, and could not find a way. *We got married with the idea of having four children and I could not even have one.* (emphasis added)

> Mariangela comments on her husband's reaction:
> He came home and did not speak for ages. I thought something serious happened at work, I thought he had lost a case for a client. He would not speak. For a moment I even thought he wanted to tell me he was in love with another woman. I did not think about the results of the test because at that point we were not that concerned, you never imagine such a thing.

During the night he started crying. I could not believe it, my husband crying! He never did in ten years we have been together. He cried and would not stop. It is at that point that I understood it was really serious, I thought it was something we could not solve in any way. I thought it was a medical thing, and immediately I realized we could not have children. He did not need to say it.

With feelings of loss and disillusion come feelings of rage and anger too. Marina (married to Michele):

I swore, yes I did. I was walking and started to talk to myself and I swore. I am Catholic, at that time of my life I was a practising Catholic, I was supposed to accept God's will. Instead I thought, 'F... off God, you have betrayed me'. I should not have said that, but I did. When I calmed down I felt completely lost. I started crying, I caught the bus home and cried all the way. I could not stop crying and felt so incredibly sad.[11]

Infertility is so shocking because it disrupts assumptions and taken-for-granted expectations. A taken-for-granted life event – settling down, starting a family, having children *with* and *from* the partner couples say they have chosen for life – is over. Infertility generates a sense of failure in the union, in the relationship, in the marriage. 'A marriage without children' Cristina drastically points out 'is not really a good one'.[12] Couples feel they have wasted time, years and years in planning the

---

11. The feeling of devastation is also widely reported by couples attending programmes with their own gametes (who, of course, precisely because of that have not yet received a diagnosis of irreversible infertility) and indeed by couples who have adopted or are in the process of adopting. Couples without vested interest say that they are able to 'imagine what it must be like' – particularly those who have already children. However, although a deeply sympathetic statement, it seems to me that it is impossible to really know what it must be like unless one has experienced a similar diagnosis. As will emerge, this work brings to light how these couples respond to such diagnosis and intense emotional pain, and what cultural models they use to elaborate their experiences upon.
12. This view strongly contrasts with that of infertile couples who are attending programmes with their own gametes and couples who have adopted or are in the process of doing so. It also differs from the views of couples without vested interest. It appears that for them, the ultimate aim of a marriage is not only a child but love, care and mutual support – although a child may be very much desired and wanted. Of course, one has to keep in mind that couples who are still attending programmes with their own gametes are in a much 'softer' state of mind because they are not yet facing the worst, and are still full of hope. Couples who have adopted, or are in the process of adopting, have gone through a period where they have reconceptualized their own place in the relationship and have invested in that to initiate a new adventure as adoptive parents. Couples without vested interest again speak in hypothetical terms and only a few of them envisage that a marriage without children would be devalued as such. It also needs to be pointed out

impossible, dreaming about what will now never take place *as imagined*. The future suddenly becomes very uncertain. The relationship vacillates and, from this moment onwards, will have to sustain extreme strain. Becker rightly asks: 'What, precisely, is at stake for them? The list is long, but at the top are having a family, charting the direction of their lives ... the challenges are profound ... maintaining a solid relationship with one's own partner' (2000: 7).

Choosing a Programme of Gamete Donation

Couples who choose a programme of gamete donation as a result of irreversible infertility do so following a rationale that often pushes them to think 'compartmentally' and that, at first, seems to contradict other views that they seem to hold. In the eyes of these couples, gamete donation is the best possible option after *natural* conception as it makes it possible to recreate what, with the diagnosis of infertility, seems dramatically lost. It makes it possible to resemble most closely what infertility has made impossible to achieve by other means. It allows the desire for a family with at least one child to be fulfilled. As if in a family portrait, it allows couples to see themselves with their (*quasi*) own child – reconfirming Stolke's point that 'NRTs are ... informed by the wish for a child of "their own"' (1988: 8; a point made also by Haimes 1990). Gamete donation allows couples to tell others the story that everybody always wants to tell and expects to hear: 'this is my baby'.

Michela (married to Daniele):
We have chosen *l'eterologa* [artificial insemination by donor][13] because with it we can have our *own* baby, isn't that right? (emphasis added)

---

that, at different points during interviews, when asked what they expected from their marriages, almost all couples without vested interest said 'to have a child'. The two statements are not necessarily in contradiction since the latter refers to an expectation, the former to what they envisage if they were to suffer from irreversible infertility – nonetheless it is significant data in itself.

13. *Inseminazione eterologa* (heterologous insemination), which stands for insemination with semen other than husband's, is the way artificial insemination by donor (AID) is addressed in Italian medical language. It is also used in Spanish and German. Stolcke suggests that it is misleading as 'in biology "heterologous" fertilisation means that different species are involved in the act of fertilisation!' (1988: 12, original emphasis). In Italian it is often used in the abbreviation '*l'eterologa*'. I do not want to be speculative just for the sake of it, but from my experience, in Italian, heterologous is a 'difficult word'. Many couples, when asked, are unsure about what it stands for. In addition, *inseminazione eterologa* does not semantically recall sperm donation, as it does not suggest the presence of a donor, of a male donor. It is not a self-explanatory term, but rather an obtuse way to describe the practice. The case of egg donation is different. *Ovo-donazione* (donation of ova) does semantically suggest the practice of egg donation. I find this inguistic

There are a number of common reasons that couples always use to justify the choice of a programme of gamete donation which strongly echo those offered by couples choosing surrogacy. Gamete donation is primarily said to preserve the *biogenetic tie* with one parent: it allows the couple to pass on 50 per cent of their biogenetic connections from the fertile partner onto the child. Ragoné makes the point clearly: 'couples who choose surrogacy and couples who choose DI [donor insemination] offer the same explanation for the choice: to have a child who is genetically related to at least one member of the couple' (1994: 114).

> Nadina (married to Ottavio):
>
> [Egg donation] gives you the chance to save what you can [*salvare il salvabile*] – it is a way not to throw away what you are left with. *Ottavio is keen to have a baby that at least comes from him,* and we thought that he should not be denied that possibility just because I cannot. The egg donor is a solution in the middle. (emphasis added)
>
> Gioia (married to Leopoldo):
>
> It is the only way to keep the family links – otherwise we loose everything. *We already had to accept that the child* will be cut off from my family in that biological sense. I do not see what else we could do. It is a trap. *At least in this way we keep his family in and the children will be connected to that side of the family.* (emphasis added)

Gamete donation also allows the experience of a pregnancy, as noted by Franklin. She writes: '*Pregnancy itself was an experience many women considered necessary to complete their identity as women*' (1997: 138, emphasis added). It enables couples to accomplish parenthood with a natural birth. As couples often put it, it allows a woman to feel like any other woman, and a man to feel like any other man.

> Costanza (married to Antonio):
>
> With an *eterologa*, I can have my pregnancy. I want to experience that feeling of being big and fat. I like pregnant women. I think they are beautiful. A pregnancy is a reward. I also think that in this way my husband will feel better. Now in the factory people make jokes because we have no children.
>
> Nadina explains again:
>
> It also gives me the experience of having a baby and of a pregnancy. That is a big thing, I think, given the circumstances. And I do not have other options. I do not want to miss out on this – I feel this is important for me in

---

clash interesting: it seems that it was felt necessary to medicalize the language of AID to the point of concealing its content more than that of egg donation. Following Hobsbawm and Ranger (1983), one could say that the naming has been *deliberate*.

my life. I deeply despise feeling different in that way, I think this is something that is important for every woman really.

Veronica (married to Giacomo):

My husband wants his own child. He gets the genetic bit, I get the pregnancy. It is better than nothing, and *it sounds much more normal.* (emphasis added)

Antonio (married to Costanza):

L'eterologa makes my wife a normal woman, and makes me a normal man, with our *own* child.

Couples choose a programme of gamete donation because it allows them to operate within the boundaries of, what they call, 'normality' or 'the normal' (quite ironically, as a programme of gamete donation also sanctions the intrusion of third parties, contradicting other cultural norms; see Ragoné 1996).[14] Couples insist on the importance of experiencing the same as everyone else, again of feeling like any other woman and any other man, and part of an imagined community (Anderson 1991) of others who never had to cope and come to terms with the drama of infertility, but on the contrary were given the privilege of experiencing what they longed for.

Such a rationale is diametrically opposite to other alternatives. The most obvious one is adoption – chosen by those couples who equally suffer from impaired infertility but who, on the contrary, when told that there is nothing else that can be done but to rely on gamete donation decide to abandon any medical treatment.[15] Indeed, for couples who choose gamete donation adoption definitely cannot work, as the two rest on very different premises and worldviews. Couples who choose gamete donation feel deeply uncomfortable with adoption. In their view it publicly exposes a family difference: the lack of biological ties between parents and children. And that is precisely what they object to and so crucially despise. They do

---

14. Although this aspect will be developed in the following chapter, it is worth mentioning Ragoné's point here. She writes: 'While genetic relatedness is clearly one of the primary motivations of couples' choice of surrogate motherhood, this view is something of a simplification unless one also acknowledges that surrogacy contradicts several cultural norms, not that least of which it involves procreation outside marriage' (1996: 362).
15. The comparison with adoption is obvious. Couples who decide to adopt find the option of gamete donation too extreme. They tend to articulate their reasons against gamete donation in the light of the fact that it is not a solution to the lack of biological ties; it is a replacement of little attraction since neither infertility nor lack of ties is reversed. They see adoption as a socially and culturally accepted practice; they often report that by adopting they do something good for a parentless child. They emphasize that, given the circumstances, they will become parents on different premises; they attribute great importance to social ties and point out that as partners they will be in an equal position towards the adopted child.

not wish to expose themselves and any family member to such difference. On the contrary they want to conform to the norm, to what they describe – and which so clearly and powerfully emerges in the quotes above – as being normal and feeling like everybody else. Couples choosing a programme of gamete donation profoundly wish to achieve conformity.

In this light, such a well-known practice as adoption, although widely perceived as a positive act, is represented as a difficult and most undesirable option. Here are some examples of how negatively these couples come to frame adoption, to the point that the *real* level of engagement and experience they have of it is often uncertain. Their take on it is so negative that it appears more of a stand than anything else.

Massimo (married to Marta):

Adoption is too difficult a process. People working in adoption are violent with families, they ask too many questions, and they are too invasive – above all social workers.

Davide (married to Lucia) already has two children by artificial insemination by donor:

It is incredibly difficult to adopt a child. Only if you are well-off they will give you a child.[16]

Teodoro (married to Olga):

We have tried adoption, but it was not successful. It is a terrifying experience, people working in there are buyers and sellers, they exploit people, they are pseudo-idealists. We have done all we had to do and got the 'clearance' [*idoneità*] for two children, but the agencies for adoption are really bad. There are too many interests in adoption.[17]

Vittorio (married to Matilde):

[*L'eterologa*] is much simpler than adoption, costs less and gives much less problems.[18]

---

16. During the interview neither Lucia nor Davide missed any opportunity to stress their wealth, yet they then claimed that one needed money to adopt. It seemed that Lucia and Davide never started the process for adoption. Later on in the interview, they pointed out that with adoption 'everybody would know'. Adoption would be an admission of infertility; conversely, with gamete donation they can keep their infertility secret.
17. In this case neither wife nor husband explained what had happened. As I experienced with other couples, it did not sound as if they really went through the system, but more as if they had acquired bits of information and then gave up. For example, Teodoro was very confused about how one gets the 'clearance'.
18. The view of artificial insemination as less expensive than adoption is inconsistent. It is part of a repertoire of commonplaces many couples seem to rely upon.

As in the case of gamete donation, and the rationale that sustains it, which often sounds repetitive and compartmentalized, the rationale against adoption often seems to follow a predefined repertoire. Ready-made answers, reflecting general commonplaces, including normative assertions of what's right and wrong, good and bad, normal and abnormal, are often deployed. Couples do not appear to be speaking from their own personal perspective and experience but rather from a collective repertoire that is embedded in their choice.

Normalizing Gamete Donation

As soon as couples enter a programme of gamete donation they try to normalize it (Cussins 1998a, Thompson 2005),[19] and work hard to turn it as much as possible into a parallel practice to natural conception. What is going to be different from what couples, in other contexts, consider and characterize as the norm – how one conceives a baby and with whom – has to become as similar as possible. This is a forced process: it requires the obliteration of anything that makes it a difficult choice. Couples enter what may be called a 'pretend' state: they may progressively define it as too difficult to accept – they may say that it is the hardest thing to do, the hardest thing they have ever done, or that they expect to be doing, in their life – or they may deny any difficulties. In the space of a few sentences, couples can make very different statements.

> Mariangela (married to Luca):
>
> It is your mind that works out everything. When you come here you *feel* lost. You know that you are going to do something which is quite unusual. But then, day by day you learn that this is your life, your way to have a child. In a month or two, coming here is something that you keep updated in your diary like everything else. It becomes part of everyday life. This does not really mean that you *truly* believe it, but it helps to think this way. (emphasis added)

> Penelope (married to Andrea):
>
> There are two things that you may think. First of all that this is altogether not such an unusual way to have a child, after all there are lots of other couples that do the same. The waiting room is full of couples who are desperately trying to have a baby like you. Secondly, that, after all, what you are really just missing is sex. It is more difficult when you suddenly realise that *this is you* actually having the problem and not someone else. When you are at work for example and a colleague gets pregnant. Then it is much harder to think this way really. (emphasis added)

---

19. The notion of normalization is extensively addressed by Cussins/Thompson, who has carried out ethnographic work in clinics for infertility treatment in the United States. As she writes, '[it] incorporates both "normal" and "normative"' (1998b: 67). I will return to it in Chapter 5, which examines the relationship between couples and clinicians.

Lucia (married to Davide):

Initially I felt it was like a betrayal. After, I started to think that we are doing it together. It is the same thing as using my husband's sperm.

Daniele (married to Michela):

It is not a problem. After all it is a way like any other to have a baby. We are no more in the prehistoric era. This is a modern world. I live in this world.

When couples try to deny all difficulties some statements suddenly sound slightly inconsistent – not dissimilar to Konrad's non-IVF donor informants who produce contrasting accounts of what eggs stand for, contain, mean, represent, pass on and signify both for the recipient and for the future offspring (2005: 58–83). Elisabetta, for instance, makes an attempt to empty the egg of any significance altogether. In her first statement she denies it any active role and property, as if it was entirely neutral and insignificant material, carrier of nothingness; as if to use a donor's egg was just a technicality. In her subsequent statement, though, she betrays herself. She, like many others, expresses a deep concern which highlights an unresolved dilemma, typical of gamete donation: what does the egg (or semen) stand for? What does biogenetic material carry? Does it carry biological information, or more than that? But what is biological information anyway?

Elisabetta (married to Germano):

I would have donated my eggs if it were necessary. Emotionally speaking, to receive or to donate is exactly the same thing. To me to be using another woman's egg does not make the slightest difference.

But soon after she adds:

My only concern is what the child will look like [*la mia unica preoccupazione è a chi rassomiglierá il bambino*].[20] (emphasis added)

Couples' strategies to settle with gamete donation – at times articulating contrasting views, at times denying the donation event altogether, and at times resorting to other domains of knowledge to sustain a programme – all aim at (re)generating a plausible narrative, a normalized one. These couples desperately attempt to look for the familiar. As Stolke remarks, the NRTs enable the 'same old fatherhood' (1986).[21]

As the account of Edoardo below shows, the attempt to justify the choice of gamete donation is made in the light of a model he is familiar with. Edoardo constructs a new narrative relying heavily on a repertoire that he knows intimately. He evokes and *uses* the past; through the past he

---

20. '*Rassomigliarsi*' may indicate physical but also psychological (emotional) resemblance. When I investigated it with Elisabetta she said to imply 'everything'.
21. This is an aspect whose importance will clearly emerge in the comparative exercise with lesbian and gay couples. I ask the reader to pay attention to what seems to me the absence of innovation and creativity in the construction of a narrative that contemplates gamete donation.

validates his own experience of gamete donation and justifies his choice. Edoardo's account is an example of the kind of variations couples may suggest, but which do not seem to represent a turning point. He, like others, attempts to normalize the practice as much as possible.

Edoardo (married to Caterina):

I choose *l'eterologa* and I do not see any problem with it. There are many children that are not related to their fathers and they are not even aware of it [the argument is that this occurs because of wives' extramarital affairs]. When I was a child my grandmother would often tell stories about children in the village [in Lombardia]. It was quite common.

Edoardo shows how a case can be put to work and exploited, and how in that light a new case (such as that of artificial insemination by donor) can be validated. In a sense it is almost irrelevant that Edoardo's association between adultery and gamete donation is rather peculiar.[22] Curiously, he seems to forget the fundamental difference between the two – at least with respect to the different role that *awareness* may play in such circumstances. In cases of out-of-wedlock conception there is, at least in principle, no awareness of the resulting lack of biological relatedness. In linking adultery to gamete donation Edoardo wipes away the cultural and social controversy that such cases cause when made explicit within the couple, as well as socially. Of course 'knowing/being aware' and 'not knowing/not being aware' have certain implications for the definition of relatedness (see Strathern 1999: 64–86). However, what makes Edoardo's example relevant to our case is how he exploits a known repertoire and what he suggests to be a cultural practice of lack of biological relatedness to validate gamete donation. He ties an idea that symbolizes difference – deviation from the norm – to another idea of difference. By doing so, he *normalizes* gamete donation.

Marta similarly attempts to elaborate acceptance. Again, like Edoardo, she normalizes the overall experience of gamete donation, associating it with a past practice (breast-feeding carried out by someone other than the biological mother – by a *balia*) even though, in a sense, these are incommensurable practices. However, as in the case of Edoardo, their incommensurability has little relevance here. What's more significant is that Marta too relies on what she feels she knows best (see Strathern 1992a).

Marta (married to Massimo):

In the past, for example, women would breast-feed babies that weren't theirs. Now a woman may carry a child that is not her own. It does not really make that much difference.

---

22. It is peculiar because the notion of adultery has been used by different parties, located in different cultural contexts and geographical areas, to vehemently contest gamete donation not to sustain it! (See for instance the Warnock Report in the UK and documents issued by the Roman Catholic Church in Italy, to cite but two).

In the attempt to normalize gamete donation couples adopt different strategies, but sometimes they can also be extremely explicit, highlighting the artificiality of such processes. In these cases, the tension of programmes of gamete donation emerges in its full complexity and can be quite touching.

> Teodoro (married to Olga):
> You have to convince yourself that these children are yours. There is no other way you can possibly cope with the whole thing otherwise.

> Federica (married to Gianluca):
> I pretend it is ours. That's all. But there are moments in which this is difficult. Last Saturday I went with my husband and a few friends to a disco, *I was dancing and at a certain point I looked around and started to think that the father of my child could be any of those men*. (emphasis added)

Do It Quickly (and It Lasts Forever)

When couples opt for a programme of gamete donation, they experience maximum tension and seem to respond by *acting out*. As Franklin puts it, 'women and couples come to feel they have to try IVF' (1997: 14). They increasingly perceive time as against them, and that it cannot be wasted. In such state of mind they make most of their decisions and choices: from selecting the private clinic, to the specific programme and the clinicians that will handle their case. From this moment onwards they will frantically also keep looking for any information that they believe may help their case.[23]

> Irene (married to Alessandro):
> We have had treatments in [a public hospital] for two years. That period wasn't particularly difficult. It was a limbo situation [there they received treatment using their own gametes] when we were told that we could possibly conceive only with an *eterologa* I felt under an extreme pressure. It seemed there was no more time left.

---

23. All couples arrive at programmes of gamete donation after several years of various attempts to conceive: first of all 'naturally', then after undergoing treatment in public hospitals with their own gametes (or in a private clinic, which also offers treatment with own gametes – private clinics can offer both). At this point, they suddenly feel under great pressure to make things happen as they are now aware that in principle they could achieve conception. 'Time' seems to play a different role in surrogacy programmes. Ragoné (1994) explains that the search for the 'right' programme (and the 'right' surrogate) is a long, complex process, where couples use time not only to finalize arrangements, but also to reflect on the choice itself.

Costanza (married to Antonio):

We spoke for what seemed quite a long time. But it was from Wednesday to Sunday. The following Monday I called my doctor [a sort of GP] and asked for the name of a specialist. Two days later we met the specialist in a private clinic.

In the logic of 'do it quickly' (which of course is quite an ironic logic at this point as so many years have passed since the initial attempts to have a baby) couples go through an impressive amount of treatments and programmes. The more the programme is unsuccessful, the more they engage in a new one. Unsuccessful treatment does not lead to rest, but conversely to action – which often also means moving to a new clinic in the hope that, there, it will work better. Each time, couples start all over again with consultations, tests, and increasingly invasive examinations. Each time, an essentially similar diagnosis will be given together with a similar prediction of chances of success and of risks involved. Clinicians always engage couples in a new programme (in eighteen months fieldwork I never saw a clinician refuse treatment to a couple, and couples confirmed to me that they were never refused treatment).

Costanza (married to Antonio):

Since I started I have had nine artificial insemination [three cycles in two different clinics] and, in the last two years, three IVF treatment [in the third clinic].

This attitude towards treatment creates a pattern of repetition. An Italian gynaecologist powerfully put it this way: 'Infertile couples suffering from impaired infertility and looking for private services are itinerant couples [*coppie itineranti*] – they go in circles'. In the desperate search for a pregnancy and a baby, they invest all their energies and resources into every possible programme they believe can give them a baby. This causes endless emotional strain (and searches).

Maddalena (married to Pietro):

I have failed seven times. Each time is the same, identical excruciating pain, but each time is also somehow different. Three months ago I wasn't as depressed as usual because I thought that wasn't really the right moment. There was something wrong with me. This time I thought 'I made it' and I did not. I am exhausted, I am not so sure I have enough strength [*forza*] to keep going, but cannot stop. In no way I can stop now.

Even though couples find treatment overwhelming, a 'hell', they do not only go for it, but never have a plan about when to stop; they say they feel unable to put any limits and draw a line, possibly because, as Cussins points out, 'each cycle takes [women] back to point zero in [the] quest to get pregnant, despite the cumulative treatment process; it also resets one's hopes, frequently only to have them dashed again in a month' (1998: 91). After a while couples incorporate treatment into their life: for the sake of having a baby, couples feel able to do this and much more.

Alessandro (married to Irene):

One ends up doing something that one never even thought could possibly happen. That makes it very difficult.

Olga (married to Teodoro):

It is like being caught into something that you cannot leave. I do not think we could do anything else at this point, for example take a period off and suspend treatment and come back in a few years. It is just not possible. There is this urgency that we feel that we need to make this happen as soon as we can. The fact that it is not happening as we thought and as we were told is making it unbearable.

Nadina (married to Ottavio):

It is the most painful emotional swing that you could possibly ever experience. It is painful – when you fail it breaks your heart but then you come back and start all over again.

Her husband, Ottavio adds:

I am not sure what is *exactly* keeping us here. But we are here, and keep coming back – and it is not working – but we will come back until it works. If it does not work here, we will go somewhere else, until it works for us as well. [24]

As noted earlier, couples can take this and much more for a baby. Although they feel deeply unsettled – twisted – they keep going. They experience the most profound emotional pain, but they keep going. They do not just keep continuing on with programme after programme, engaging in a new one as soon as an IVF cycle is unsuccessful, but also go through the effort of appearing OK to the world outside; they make every possible effort to maintain a façade and put in place several strategies for survival.

Cinzia (married to Michelangelo):

We have like two stories going on: one is about us, what we know about each other, something that nobody else is aware of. The other one is about what we say to everyone else; about not having a child yet.

Michelangelo adds:

It is like living a schizophrenic life. Not sure how long we are going to be able to keep going this way. But in a way we have no choice.

---

24. Couples who undergo programmes with their own gametes and adopting couples point out their unwillingness to engage further with treatment and higher levels of medical invasiveness. These couples, contrary to couples choosing gamete donation, seem to have a clear idea of when and why to stop. They also make clear statements about 'drawing a line' before and not after engaging in a programme.

Davide (married to Lucia):

It is a bit like performing. When you came here you have to show that it is OK. When you are outside you *lie* all the time as soon as somebody asks you a question about the children that you do not have. (emphasis added)

Marta (married to Massimo):

I keep making up stories. I have always to take time off work and each time I have to invent a new disease [*disturbo*]. Surely people wonder how it is possible that there is so much wrong with me.

As time is crucial at all stages – during the process, before treatment, following treatment, whilst waiting for the results, to the moment of eventually having to re-engage in the programme – couples lead a 'suspended life' for the time that it takes, and often for a time that feels (almost) forever. As Franklin says, 'it is because the demands of IVF are so intensive that the procedure comes to feel like a "way of life"', something you "eat and drink", that you "live" twenty-four hours a day' (1997: 130).

CHAPTER 4

HETEROSEXUAL COUPLES: GAMETE DONATION, DONORS AND BIOGENETIC MAKE-UP

A Case: Matilde and Egg Donation

Matilde (married to Giorgio) is twenty-eight years old, an ex-model. Matilde, who suffers from premature menopause, says she is desperate for a child, her *own* child. She has already had IVF treatment with egg donation but failed and is now on the waiting list for a second attempt. Although she is hoping to be treated soon, she finds it difficult to accept the genetic contribution of an egg donor. If she had the choice she would rather attempt a new treatment available in the United States that, in her words, would still allow her to maintain a genetic link with her baby despite her impairment. In order to undergo such treatment (for which I was never able to find supporting information) she has to travel to the States and enter the list of volunteers for testing the new technology. She asks me for help, as if I could do more or knew more than she does, which I do not. Matilde says that she would do anything, undergo any treatment, take on any programme if she were guaranteed to be able to contribute *her own genetic material*. Matilde says that it is very important to be able to have a child that is genetically related to her, and she explains why:

> I think that a man marries you because you have that face, that body and that mind [*quella testa*] and he wants the baby to take from you. I am waiting for egg donation and have already had an IVF cycle. It failed. If the American treatment becomes available in Italy I will go for that. Even if it is still experimental. *Not because genes are really important in themselves, but because my genes are a part of me and my husband married me for what I am, he did not marry somebody else.* (emphasis added)

Matilde's predicament, like that of Anna at the start of the previous chapter, is illuminating because it throws in great relief some of the tensions that couples experience, although not all of them make them so explicit. Matilde makes it clear where the problem lies – that is, in what she thinks *she is:* 'that face, that body and that mind', the whole person, and the whole person in her physicality and personality. That really matters. She finds it difficult to split herself by accepting egg donation; she find it difficult to incorporate somebody else as she believes that all that she is, is inscribed in her biogenetic make-up (hence other people's physicality and personality are inscribed in their own biogenetic make-up). Matilde wants to keep all of herself together. In her view the whole of herself makes her special in the eyes of another like her husband and therefore she strongly does not want to give it up. At a different point in the interview Matilde makes it clear why:

> There are many women who always want to change things about themselves. They look for cosmetic surgery and want to re-make everything if they can. That's not me. I was always content with how I was born. How my parents made me. With all my defects and everything. That's me.[1]

Introduction

The previous chapter describes two different moments in the lives of couples: the moments either side of a diagnosis of infertility; the time of dreams and expectations before, and the time of compromises after, once the diagnosis of infertility is there and is irreversible. These two moments can be formalized as a shift between the ideal and the real – what in common parlance is often referred to by Italians as *l'ideale e la realtà*. This chapter explores another sort of duality. It explores couples' conceptualization of biological inheritance and biogenetic make-up (*eredità biologica* and *patrimonio genetico:* the former stands for what is transmitted, the latter for what individuals are made of), perceptions of donors and the practice of donation, and processes of acceptance of both. It shows how complex it can be to have chosen to undergo programmes of gamete donation with third-party assistance, and to formalize a view of this. Most couples keep shifting and find it difficult to make their position explicit. They adopt different strategies to familiarize themselves first, and then to cope with certain cultural complexities that programmes of gamete donation generate: if, on the one hand, such programmes allow couples to conform and *perform* certain cultural and social prescriptions, on the other

---

1. Matilde's case is particularly interesting as her state of mind seems so close to that of infertile couples opting out of gamete donation, choosing adoption and that of couples with no vested interest in assisted conception who express discomfort with programmes of gamete donation. However, Matilde, contrary to them, despises adoption or the prospect of remaining childless.

hand they also profoundly jeopardize others. Gamete donation makes it possible to emphasize the biogenetic tie between the fertile parent and the child, but at the same time it requires downplaying the lack of tie between the infertile parent and the child and the biogenetic tie between the donor and the child. With a programme of gamete donation couples need to fully revisit the notion that 'blood is thicker than water' (see Schneider 1980 [1968]; Wolfram, 1987) – the Italian notion that *il sangue non è acqua* – but in *asymmetrical* ways. Blood *is* thicker than water with reference to the biological tie that is preserved between the fertile parent and the future baby, but *it is not* with reference to the anonymous donor and the baby. This is a dissonant and complex exercise – particularly for these couples who are looking for *normativity*.

Infertile Couples, Biological Inheritance and Biogenetic Make-up

When it comes to notions of biological inheritance, biological processes, genetic make-up and the role of genes in the making of persons, couples seem to adopt different strategies to conceptualize and formalize their views, and what they believe to matter. They treat these particular issues in various and contrasting ways – especially if read in the light of other statements (see extracts in previous chapter). They, who have been most personal and emotional, now seem to crave detachment and disengagement.[2] Most couples find it easier to depersonalize statements and dissociate themselves from any personal position, knowing that it would make them feel uncomfortable. Suddenly, they tend to speak in terms of what others think rather than what they think – and although views and opinions still show through quite clearly, they make a great effort to make them impenetrable. They seek to hide the significance that they attribute to biological inheritance and biogenetic make-up, and the role played by the genetic contribution of third parties (egg and sperm donors) in the making of offspring. In other words, they put in place what I have called *processes of obliteration*.

Caterina (married to Edoardo):

Many people would say that to have biological ties [*legami biologici*] with a child means that there is a part of oneself around in the world, and that when one dies that part is going to remain. So for example things like eye colour, or voice or some aspects of one's character won't die. These things are very important for most people. It is a way to perpetuate oneself, and

---

2. The attempt to keep some distance is particularly significant here. Without stereotyping unduly, Italians tend to personalize points of view. Generally speaking, to be too impersonal carries a low value. In conversations on the role of biology and genetic make-up this space is often left empty and many couples talk in terms of what is ordinarily understood by *others*, not by them.

also it gives that feeling that one won't die completely. Because presumably children will have other children, so there will be always a continuation. *I do not think this is so important anyway.* (emphasis added)

Giacomo (married to Veronica):

Eggs and sperm make the essential part of a person. I mean each one is different from each other because that particular mix of egg and sperm. For that reason it is important for people. It seems to me that *people* also think that genes make you intelligent or stupid – because it is a question of DNA. *People think* it is written, what you are, and what you are going to be. If that part is not very good you may have fewer chances in life. There are also many other things that count though. (emphasis added)

Leopoldo (married to Gioia):

It depends. *Some people* think that the biological tie [*legame biologico*] is everything and also that genes [*geni*] are the building blocks. Others think that it is more down to the environment. *People think* all sorts of things but no one really knows the truth, although *most people* also think that we inherit from our parents our physical traits and also our personality. So I suppose that's the main view. *Most people* would say that. (emphasis added)

In the logic of detachment and disengagement some couples adopt an even more radical strategy: they claim that biological inheritance and biogenetic make-up have no influence and play almost no role in the making of individuals. They fully and uniquely emphasize the role of the so-called 'environment' (*l'ambiente*).

Roberto (married to Letizia):

I am not sure I entirely understand why *some people* are so obsessed with the idea of the blood tie [*legame di sangue*]. They surely believe that everything starts from the genes [*geni*], that the genetic make-up [*patrimonio genetico*] contains everything, that it is all written, but it is not. *It has been proved that it is all down to the environment and the genetic make-up has nothing to do with us.* (emphasis added)

Vittorio (married to Matilde):

Everyone knows it now – psychology has shown it very clearly – that *the genetic make-up [patrimonio genetico] contains almost no information*. The child starts from zero and it is up to the parents to help out the child to form the personality and everything else. Take the example of twins. Their genetic make-up is the same but they are always very different. *It is the psychology of the family that makes it all.* (emphasis added)

Emma (married to Guido) represents an in-between case. Although she also relies heavily on the point of view of imaginary others, and like others points out that questions of biological inheritance and biogenetics are not relevant to her, she is more prone to personalize her view:

> I believe that *many people* do not know very much about biology or genetics, I guess they want their children to look like them and have something of them. *For me it is not important.* The idea of being the same, of having something in common, I think that one just makes a family, and it really does not matter if you have the same blood. *I never think about that anyway.* (emphasis added)

Stefania (married to Mattia) tackles the issues in a personal fashion too, but she also insists that questions of biological inheritance and biogenetic make-up have little relevance for them.

> We had quite a few discussions and came to the conclusion that for us this is just not really the point. *In a scale of values the biological tie is the last thing [in una scala di valori il legame biologico è l'ultima cosa].*[3] (emphasis added)

In all these cases, when couples either completely depersonalize statements, talking through others or take a position in-between, reaching the conclusion that biological inheritance and biogenetic make-up have little or no relevance, one cannot but wonder why they feel the need to make such drastic statements. Such statements are made all the more incongruous by the efforts that they make to go through programmes of gamete donation, which have been chosen primarily to preserve 50 per cent of the biological tie with their offspring. They are clearly in utter contradiction of the very essence of a choice of a programme of gamete donation.

Such incongruity becomes even more evident from the accounts of couples who, on the contrary, feel able to personalize their views fully. They highlight, often by implication, how painful the loss of biogenetic connections can be. Such revelations point to the place biogenetics occupies in people's conceptualizations of ties and to the hurtful effect that dwelling on it can have. These are revelations that expose what detachment, disengagement and *processes of obliteration* conceal.

---

3. As already pointed out, the above quotes seek to convey the relevance that such matters have for others and downplay the relevance that they may have for the couples concerned. Interestingly, the data I happen to collect with *others* – once again couples attending programmes with their own gametes, choosing adoption and couples without vested interest – say something different. *Others* seem to construct their argument on different premises altogether: those who believe that biological ties and biogenetic make-up are indeed very relevant in the making of individuals, in the construction of personal identity, in forging emotional and social relations and so on claim that gamete donation is unacceptable precisely because it introduces the genetic material of third parties who are strangers to the intentional couple and therefore carriers of the unknown and unpredictable. On the contrary, those who believe that biological ties and biogenetic make-up are indeed not that relevant in themselves do not see a plausible justification for choosing gamete donation in order to preserve 50 per cent of the biological tie and biogenetic make-up between the fertile parent and the child. In all cases, thus, they see gamete donation as a fundamentally unsatisfactory option, which profoundly jeopardizes the cultural norm.

Sonia (married to Eugenio):

I think that genes means that one comes from the other. So that *my children come from me. It is something quite special between people, because not everyone has the same with everyone else. It makes my children special to me,* it also makes them special to me and my husband, because *they come from both of us together.* (emphasis added)

Federica (married to Gianluca):

The blood tie makes your children very special. But it is more about knowing that they belong to you and vice-versa. It is important because they will be able to trace everything back. It is a line that links everyone in the family. *It is the history of the family too.* (emphasis added)

Marina (married to Michele):

I think that the biological tie [il legame biologico] is important between a mother and a child. It is important to me, as a woman and as a mother... and it is important for the baby. *It is important to know that you are connected, the same to each other [è importante sapere che sei legato, uguale l'uno all'altro].* It makes me feel very close to my family and it makes things easier, it makes me feel more secure to know where things come from. We all need to have a tie.

Davide (married to Lucia):

It is hard the idea that *I am dead.* No matter how many children I may have, none of them will have a bit [*una briciola*] of me – of my blood. They will not have bits of the family – there is no history left there – built up through generations.

Statements like those above, and particularly like that of Davide, highlight how much there can be at stake for couples. Davide feels he has lost the reflection of himself in his children and that connection that makes the father-child tie precious precisely because it is unique and irreproducible in others. Davide sees the biogenetic tie as continuity and perpetuity; its absence makes him feel that he has lost his own place in the world and that his own existence will come to an end. Infertility makes it impossible to regenerate the 'gene pool' or 'blood line' (*la stripe, lo stesso sangue*), which is more than a loss of genes (expressed in terms of 'I am dead'). Expressions like 'me', 'bits of the family', 'history' stand for the whole person in a past and present social context. Edwards also traces the common idiom of the 'blood line' in conversations with her informants which, as she rightly points out, might be a 'genetic connection, but it goes further even as it embraces it' (2000: 217).

Overall, couples look for ways to rationalize the anxieties that questions of biological inheritance and biogenetic make-up generate. They seek to harmonize in various ways, with the aid of different strategies, powerful dilemmas. Strategies of detachment and of, what I call *obliteration*, whereby couples go as far as they can to wipe out any relevance they may otherwise attribute to biological inheritance and biogenetic make-up, clearly show the broader politics of denial. More open statements, on the contrary, show how disturbing it can be to hold certain conceptions of biological inheritance and biogenetics when there is full awareness that in one's own case, and due to the choice to undergo a programme of gamete donation, there is much to compromise on.

Couples' Perception of Donors and Donation

Couples are fully aware that gamete donation is not a medical intervention in the strict sense.[4] Gamete donation is a procedure, often highly complex and technological,[5] that requires the aid of clinicians and medical staff, but most of all it rests on the participation of others – egg and sperm donors – to achieve its objective. Without donors there would be no intervention and no treatment. It is thus one of those interventions that carry cultural and social complications because of the involvement of third parties (and the act of donation itself).

The figure of the donor occupies a particular place (see Daniels and Haimes 1998). The donor, with the act of donation, makes her/himself present, but is also anonymous, mysterious and unrevealed in any other sense. Couples have no conception or *real* information about donors – who they are, where they come from, their background, what they do, how they look, and most of all their motivation for donating (see Daniels 1998).[6] Couples undergoing treatment have that reference in mind, a formal knowledge of that presence, but little information to rely upon and to work with. This lack partly contributes to their feeling unsettled, partly pushes them into a corner where they feel unable to ask for clarification, and partly (most importantly) it contributes to stoking their imagination, producing different visions of donors and the practice of donation as a whole. As Lasker points out: 'perhaps the greatest concern ... is over the identity, the health and the character of the donor' (1998: 10).

---

4. Gamete donation is not always performed on the body affected by a pathology; on the contrary, it is frequently performed on the body unaffected by it. Additionally, it is not a medical intervention that remedies infertility but simply allows access to the fertility of others, as Strathern has pointed out (1990 – see Chapter 5).
5. Of course it depends on the treatment. Artificial insemination by donor is a simple procedure that can be performed in a DIY fashion (see Chapter 6 on lesbian and gay couples) – it requires only skill and instrumentation for the conservation of sperm; IVF is a much more sophisticated technology, particularly if accompanied by other medical and surgical procedures.
6. The provision of minimal information is in line with most European countries' practices up to the late 1990s (see Glover Report 1989 and further notes later on). It is in contrast with North American attitudes where couples can increasingly 'choose' their donors.

Francesca (married to Marco):

You always have that side-thought in mind. You think about it, even if you don't want. It comes into your mind, when you work, or eat, or speak. It is sudden, it may come in obvious ways like if you see a child or a pregnant woman, or in less obvious ways. If you see a spotty guy in the underground and you think awful things like 'I would not want that guy for my child or for my donor.'

Michelangelo (married to Cinzia):

It is not a deliberate thinking; it just happens. You cannot really help yourself because there are too many situations that make you think about it.

Margherita (married to Pier):

The donor is there, and you are aware of it, because if you are ever going to become pregnant it is because you are doing an *eterologa*, but it is hard, you need to find ways to cope with it. You try to concentrate on other things, for example the doctors, who was there this or that time [doctors take it in turns] or what doctor X did or said. Of course in the back of your mind you are thinking other things. *It is impossible to forget the donor.* (emphasis added)[7]

The donor is often there, in everyday thinking about the treatment, the pregnancy and the baby. Prior to achieving a pregnancy, all couples admit to thinking about the donor all the time, asking themselves questions, to be having doubts, wanting to know, but feeling too intimidated to ask. Lasker identifies a few studies where it emerges that Polish and Danish couples would like to have more information about the characteristics of donors, and most of all express the wish that they (donors, in cases of DI) look like the husband (1998: 11).

Marina (married to Michele):

I would like to understand and to know who these people are, I have so many questions to ask, but I cannot.

Penelope (married to Andrea):

It is like a black wall. I know that there is somebody there behind, but I cannot see it.

Giulia (married to Fabrizio):

It is painful – more than once I found myself dreaming princes and kings [*mi sono ritrovata a sognare principi e re*]. Other times I have tried to be more rational.

---

7. Margherita was one of my main informants during fieldwork; her way of putting things seemed to break the ice easily. She had the ability to say not only what she thought, but I suspect what other couples, and I too, would have wanted to say or were perhaps thinking of but for various reasons found difficult to make explicit. On some occasions I used her comments to stimulate discussions with other couples.

Matilde (married to Vittorio):

It is psychologically devastating to think of this person that I know nothing about. When it comes to my mind I just ignore it and I think things like *at the end of the day who is the donor? Just a bunch of spermatozoa, nothing else.* (emphasis added)[8]

These statements all testify that donor and donation are not without attached meanings as Novaes Bateman points out. She writes: 'donation is not a purely physical act; semen, oocytes, embryos, pregnancy and children are all strongly charged with meaning ...' (1989: 642). In different ways, couples attempt to come to terms with the idea of the donor and donation: sometimes desperately wanting to know, sometimes preferring not to.

Most couples attempt to suppress such uncertainties and anxieties, projecting positive images as much as they can. They seem to need to think about donors as individuals who have a mission in life to help infertile couples, without expecting a reward. They conceptualize the donation of gametes as a pure and unpaid act. They imagine donors as generous, understanding and humane (*persone umane*).[9] They imagine the kind of non-IVF donors described by Konrad in the UK who, amongst other reasons, are motivated by a desire to give the gift of life and the joy of a baby to childless couples (2005: 61–73). They produce a rhetoric that is produced elsewhere and that sustains programmes of ova donation in different contexts, under different rules. Many couples go even further, and fully assimilate it to the idea of blood and organ donation, which in their eyes have a strongly positive cultural and social image. They categorically do not

---

8. All these quotes, of course, show a tension. Couples who attend programmes with their own gametes, couples who adopt and couples without vested interest all find third parties' contribution extremely problematic. The absence of any contact, information and relation with the donor is perceived as 'highly disturbing' as many put it. Most couples see third parties as carriers of 'traits and personality' that would make themselves present in various ways and forms in the child and that would be unpredictable. Many see also that such 'traits and personality' would combine themselves with those of the fertile partner and that would evidently exclude the husband/wife who suffers from infertility. They envisage such a prospect as creating great imbalance within the couple. In addition, these couples produce quite realistic images about donors and the practice of donation. As Francesca says, 'It is difficult to rely on other peoples' [donors]... that one does not even know who they are [*è gente che uno non sa chi sono*]'. Adopting mothers often use the expression of somebody else's sperm inside as 'disgusting'; they say the same of donated eggs. Marta, for instance, says: 'I think that my body would just reject immediately another woman's egg ... that's totally crazy! [*È una pazzia!*]'. In the British context, Alltown women seem to find the idea of gametes from an anonymous donor quite problematic too (Edwards 1999 [1993]: 77).
9. The expression 'persona umana' (human person) is often used by couples. It is quite meaningful in Italian; it expresses the maximum capacity of a person to be deeply involved and concerned for somebody else.

want to reflect on the substantial differences, including the fact that it is severe illnesses and survival which encourages the donation of blood and organs internationally and which raises a different order of ethical issues (see Scheper-Hughes and Wacquant 2002).[10] Here are some images.

Giacomo (married to Veronica):
Donors do so because they want to help people like us. They just *give*, they do not *take* anything. (emphasis added)

Edoardo (married to Caterina):
Donors do it for the same reasons they would do blood donation.

Caterina adds:
Alike in all other cases of donation, they want to help suffering couples, also this makes them feel that they have done something important in their life. They have helped. They have donated happiness, the joy to live again to couples like us.

Cinzia (married to Michelangelo):
I read that donors are very compassionate people, and they may be doctors themselves, or may have read about the suffering of infertile couples and decide to donate. I do not think there is a great difference between egg and sperm donors. They want to help and get in touch with clinics and make a donation. I think this is how it works, I do not think they get paid or anything. Perhaps they get their expenses like petrol or the bus ticket reimbursed.

Guido (married to Emma):
They are very young medical students, that is what the doctors told us. Everyone says that, that must be the case. They study infertility and understand what it means so they make a donation I think. It is a bit strange to put it like that though because with all the things that they study if they became emotionally involved with everything they would have no body parts left. It is a bit strange but I am pretty sure that the donors in this clinic are medical students.

In contrast, the extracts below, show how some couples (not surprisingly, a small minority as these are unsettling questions and most couples avoid them) ask themselves more pressing and disturbing questions: Do donors really *donate* to the clinic egg and semen for infertile couples? Are they really

---

10. In contrast, couples who attend programmes with their own gametes, couples who adopt and couples without vested interest see a substantial difference between the practice of blood and organ donation and egg/sperm donation. They can clearly see the difference between the former as a practice for human survival, and the latter on the contrary as a practice that creates lives anew. These couples often think about adoption as the alternative solution to infertility and believe that, as many say, 'with it you also help a parentless child'.

compassionate individuals as everyone portrays them? What are they really like? Why are they doing it? Do they get rewarded for the donation?

Margherita, who is especially explicit, puts the matter in clear terms:

> If donation means that there are some men *selling samples* of their sperm why do we call it donation? It is wrong. (emphasis added)

Pier, her husband, adds:

> I would not accept it, if it was not a donation. A person that donates, only donates – that must be the case. It must be like blood or organ donation.

Fabrizio (married to Giulia):

> I would find it very disturbing to discover in a few years' time that donors were just people who did this for money. I am sure that's not the case, if that was the case doctors should say it. I hope so.

Pressing questions about donors and the act of donation also come with equally pressing questions about the chain of actions that lead to the procurement of gametes. Through specific actions couples try to imagine those to whom the gametes belong. It is as if couples, in locating the act of donation, attempt to put to rest questions that keep coming back, and can construct a version of a narrative about donors, donation and gametes. I say 'a version' because couples never see, and will never be able to see, the transfer of gametes that eventually make possible the conception of their baby.[11]

Fabrizio (married to Giulia):

> I wonder where they put the semen, *I mean where are all these potential children to be born?* Are they in stand-by? I would pay a fortune just to see once forever *where they stay*. (emphasis added)[12]

---

11. Whilst, as mentioned, very few couples attending programmes of gamete donation are able to be explicit, those not involved in the programmes are fascinated by how clinics work, and particularly by how the transfer of gametes occurs within clinics. Couples who refuse gamete donation and couples without vested interest also pointed out this aspect to me. Indeed, these were also questions frequently asked when I was interviewed by the media during fieldwork (e.g. *Radio Capital*, Rome 1998; *La Repubblica*, Rome 1998).
12. This is one of those statements that struck me. What Fabrizio would like to see is what I believe has helped me to understand gamete donation. Between one patient and another I would spent time in the lab. There I would take further notes or I would chat with one of the biologists and ask questions; or would just observe what they were doing. Most of the time they would be working on donor sperm before freezing it into big cold (– 196°) cream colour containers. In another clinic where they performed surgery I observed the procedure of collecting eggs as well as fertilization and conservation of pre-embryos. I believe containers, fridges, needles, syringes and so forth *spoke to me*. I soon realized what a privilege it was to

Sonia (married to Eugenio):

> If they let us see, it would help I think – at least where they keep the eggs, how they keep them, how they treat them, you could name them, make them yours for the pre-implantation period at least. It is all a bit obscure. They do not do that because they do not belong to you, they are still donors' eggs. Although you pay for every bit of treatment. They belong to the donor. Or to the clinic perhaps, until they are implanted but even then they are still donors'. It is pretty awful to think in these terms.[13]

This should be not so surprising. As the donation of egg and sperm is strongly charged, couples attempt to locate both the donor and the act of donation in a familiar realm that, once again, allows them to come to terms as much as possible with the agent and the practice as a whole. Strathern also writes: 'Donation is ... [can be] conceptualized in two ways. On the one hand it may simply involve an act of bodily emission intended for an anonymous recipient; on the other hand it may involve a *relationship* between donors and recipients as partners in a single enterprise' (1992b: 128, emphasis added). Couples, by producing positive images about donors and their motivations, and by associating it with a practice that they have some familiarity with such as blood and organ donation, seek ways of reconciling anxieties, fears, uncertainties, contrasting feelings and thoughts. When they question it, they do so in the hope of finding affirmative, positive answers.

Good Intentions, Gifts and Donors' Displacement

As we have seen, prior to achieving a pregnancy, couples most often construe an ideal image of donors. Donors are imagined as people committed to the cause of infertility, willing to help those incapable of autonomous procreative capacity in the name of a special sensitivity. They are mostly believed 'to be able to feel' what it must be like not to be able to have one's own baby. They are thus believed to be motivated by a moral call, and never rewarded for the donation. Despite pressing doubts (and fears), couples always locate the donation of gametes in the arena of the 'gift'. Gamete donation is considered equivalent to blood donation and even to the donation of organs: '*è un dono*', couples say. The idea of the *gift*, which is instrumental to the idea of an *altruistic act*, evokes positive images/associations about the moral status of the person who donates.

---

be able to simultaneously experience the lab setting and couples' accounts. It was a privilege that couples who undergo treatment are not allowed and that Fabrizio here is unwarily addressing.

13. The naming of gametes and their ownership status resonates strongly with the comments of Edwards' informants, who express a discomfort with clinics 'owing' gametes and embryos. Whilst embryos do belong to couples and enjoy much protection, gametes enjoy a much less clear status. This similarity of concerns is quite striking as quite specific (2000: 233).

Strathern writes about Euro-Americans: '... to Euro-Americans, gift-giving seems a highly personalized form of transaction' (1992b: 122). The donor, in donating eggs and sperm, is believed to donate a part of her/himself in the view of couples (see also Konrad 1998, 2005). The moral status of the donor – perceived both as the provider of biological material and an active participant in the conception of a very precious child – is of paramount importance. Such status is a reassurance – almost a *guarantee* – for the future; his/her moral qualities lessen the anxiety experienced by couples. How donors are dealt with changes through time, however. Even if at the beginning, from the time of engaging with a programme to the time prior to achieving a pregnancy, couples keenly produce positive images and are (in different degrees) able to talk about donors, their role and can produce a variety of images and account for their existence, as soon as treatment progresses – and certainly as soon as they achieve conception – they *feel* the need to move away. Couples increasingly create distance and depersonalize donors: they turn donors from persons of high moral status, motivated by very good intentions, into body parts, namely body cells. Persons, in their wholeness, dissolve. The transformation happens strategically. Almost proportionally to the length of treatment, donors are increasingly less and less mentioned. Their role is de-emphasized and minimized until images of total irrelevance are produced. The donated gametes are gradually treated as belonging to non-persons. They are treated as if overall they play an incidental role. Couples (heavily encouraged by clinicians) never talk of donors' gametes; instead reference is always made to other aspects: the woman's reproductive body, the quality of the partners' (wife/husband) sperm and eggs and the quality of the embryo as if it 'consisted of' the genetic material of the fertile partner only.

> Francesca (married to Marco):
> You do realize after a while that you are the main player in the overall thing. It is a bit as if you were having this child by yourself.

> Margherita (married to Pier):
> The donor slowly fades away. The doctors help you tremendously in that. Slowly somehow he is out of the scene on a day-by-day level – from being an overwhelming figure, he is suddenly no longer around as he used to be. You get used to the idea of it, I suppose – a bit of a brainwash.

> Then she adds:
> At the end of the day what really matters is the baby and I have got used to the idea. I stopped thinking about everything else and started concentrating on the baby. I thought that if I put all my energies on the baby I would get pregnant and as soon as I started thinking this way I actually got pregnant for the first time, then I lost the baby, but since then I just stopped thinking about everything thing else and stopped being obsessed about which donor the doctor would assign to me.

Once conception is achieved donors are definitely never mentioned again. The couple carries on with the successful pregnancy and never refers again to the donation of gametes. The idea is that a donor was present (and necessary) at a point in time only. As Novaes says of French experience, 'minimising the donor's contribution and eliminating all traces of his participation have always seemed a positive attitude permitting the couple eventually to forget the procedure and lead a normal family life' (1989: 640). The presence of donors is thus fixed to a point in time in a couple' life. Thereafter, it disappears forever. Such a strategic device leads to a politics of secrecy.[14] The displacement of the donor in time – as soon as treatment is undertaken and certainly as soon as conception is achieved and from that moment onwards – is fully and powerfully instrumental. There would be no point in putting in place a strategy of displacement if the ultimate aim was not secrecy.

Sonia (married to Eugenio):
We have discussed it several times and reached the conclusion, after talking to the doctor too, that it would only be harmful to the baby and the family to tell the truth. This is a kind of truth that would only cause pain and harm. The baby may feel unwanted and *may be looking for the real mother forever.* This would be too painful for everyone to bear. (emphasis added)

Letizia (married to Roberto):
The doctor said to say nothing, that there is no point. I think he is right. This is already the most difficult choice we could make. The baby will not be able to get in touch with the donor anyway and that would be dreadful for us. It is much better to keep the whole thing secret and that's it.

Gianluca (married to Federica):
We are definitely not going to say a word about this to anyone.[15]

---

14. Couples undergoing gamete donation never confide with others that they are doing so. They never say that they are relying on egg and sperm donors to achieve conception. They are highly supported in this by clinicians, who reinforce the idea that gamete donation should be kept between the couple and never communicated to the baby to be born. In Europe and America the trend has been very similar (see Haimes 1992 for an earlier discussion; Daniels and Haimes 1998; see also Frith for an overview, 2001). More recently there has been a shift towards 'transparency' (see Strathern 1999; Konrad 2005).
15. Several studies show that couples, even in contexts where donors are non-anonymous such as in Sweden, tend not to tell children that the means of conception involved donated gametes. In Sweden, a study found that 89 per cent of couples had not informed their children (Gottlieb et al. 2000). A Dutch study found that 74 per cent of DI parents did not tell the child either (Brewayeya et al. 1997), and 73 per cent in the United States (Klock et al. 1994 – for an overview see Frith 2001). This is in complete contrast with the view of couples who attend programmes with their own gametes, couples

Besides revealing couples' attitude towards secrecy and transparency, and clearly illuminating the policy of displacement, these statements, once again, are illuminating because they reveal from another focal point specific cultural constructs; foremost, the notion that secrecy needs to be pursued in order to prevent the possibility that one's own offspring may find the urge to look for the 'real' mother (or for that matter for the 'real' father) at some time in the future. It is thus implicit in many quotes that biological connections may be paramount and, no matter what, the desire to look for them may prevail in the offspring. Couples seem unable to account for what they will have put into the relationship, and once again emphasize mere biological ties. They portray biological ties as catalytic: they project the fear that offspring will act on them and in their name activate a chain of actions to look for what couples name as the 'real' parents (which in any case would be impossible in the current legislative climate). They fear that such a chain of actions may lead to new forms of relations. They envisage a similar situation to that of 'open' adoption and dread it tremendously. As Strathern points out 'today's problems are the ("natural") parents. For the ("social") child is bound to want to know, it is said, what its biological antecedents really are' (1992a: 53). This is an awkward prospect for couples to predict as – at least in principle, from a biological perspective alone – they are in the same position as the providers of biological material! They share with their future offspring 50 per cent of biogenetic make-up. But, it appears, they find it difficult to think in those terms, just as they seem to find it difficult to think consistently that, beyond the biogenetic tie, they will have built over time powerful social relations, powerful ties of love and care. Although at a formal level, it is precisely that awareness that enables couples to choose programmes of gamete donation, somehow it gets overridden and becomes, so to speak, insufficient to suppress anxieties when the possibility that the biological provider may become known exists. That prospect alone triggers much apprehension and concern, because as Strathern points out for the English, '[it is] when persons become visible as individuals that ... they "relate" to one another' (1992a: 49).

---

who adopt and couples without vested interest. They unanimously and unequivocally find the idea of forbidding a child 'to know where he/she comes from' quite problematic. Irene, a mother of two with no vested interest in assisted conception, says: 'It is a sort of lie that I would not want to have to say – it is a 'right' [*diritto*] that I would not want to deny to my children. I am not sure how they [people] cope with that sort of secret in the family, I would be quite concerned about when they [children] became adults'. Interestingly, this strongly resonates with a comment made by Dame Mary Warnock: 'I cannot argue that children who are told of their origins, if they are AID children are necessarily happier, or better off in any way that can be estimated. But I do believe that if they are not told they are being wrongly treated (1985: 151 cit. Firth 2001: 822; see also Daniels 1995; Freeman 1996 on child's right to know; see also Lasker 1998).

CHAPTER 5

HETEROSEXUAL COUPLES AND CLINICIANS: STRATEGIES IN PRIVATE CLINICS OF ASSISTED CONCEPTION

Extracts from Field Notes: at Lunch with Clinicians

Today I went to assist an egg extraction with the clinicians and then another lady had an embryo transfer. When they finished and we got our surgery clothes off, we went for lunch. As at some point we were all women, I thought I should ask if they could not have a baby, would they use a donor themselves? When I was wearing my green suit and my light blue cap I was thinking that there is something awkward about the whole thing and was not sure why I was thinking that –perhaps because the woman was crying and saying 'this is my seventh attempt' and I found it touching, but also somehow unacceptable and painful. So I thought I should ask the question: 'Would you choose gamete donation or would you perhaps choose adoption?' They all suddenly stopped eating. For a moment, which seemed forever, no one spoke. The *air froze, as did their faces and mouths*. I think they all looked at each other and made a few noises and then said things that were all non-answers, just noises. I should have never asked that question.

Milan, March 1998

This extract from my field notes describes a moment of heavy tension with clinicians and the fact that although, at this point, I had been attending the clinic for a while and had developed personal relationships with each of the clinicians, it was really never that clear what could be said and made explicit and what should be left unsaid. Anything that had to do with the programmes, treatment, couples, motivations and choice – but especially donors and the practice of donation itself – was a highly sensitive subject and could be discussed only by following certain rules. My question over lunch was instinctive and emotional, dictated by the circumstances and the encounter with the woman who was undergoing the procedure. As such it was unwelcome. It put clinicians on the spot. Particularly, as I learned later, no clinicians would readily choose a programme of gamete donation if suffering from infertility. On the contrary, when I investigated this further with other clinicians, I came across interesting revelations: clinicians suffering from irreversible infertility often decide to adopt or to remain childless. Their work in the clinic is their *professional persona* and there they seem to buy fully the rhetoric that they themselves sell.

Introduction

The two previous chapters explored the state of mind of couples from the moment of discovering infertility to the time of choosing a programme of gamete donation, with all the implications and complications that come with that choice. They also dealt with notions of biological inheritance, biological make-up, biological processes, donors and the practice of donation, including what I have called *processes of obliteration*. This chapter deals with an additional level. First of all it frames a specific rhetoric about treatment, services and choice, which has been emphasized as a consequence of the legislative context: in the absence of a statutory law on assisted conception (which came into place only in 2004, see note in Appendix I) private clinics have been operating with no regulatory framework. Perhaps, this vacuum has also made it possible to produce opposite rhetorics such as, for instance, that of hyper-medicalization and the ill body which are in subtle contrast to that of personal choice and demand for services. The chapter examines the interactions between clinicians and couples. Programmes of gamete donation are designed, applied, managed and sold by private clinicians. In selling programmes they also sell ways of thinking about them: ideas are pulled together from different domains, the medical language is replaced with what I have called the *language of commonplaces*: accessible and recognizable by anyone approaching treatment, able to cut across class and status and to touch the core of very specific kinship notions. It is a language that suppresses anxieties, fears, doubts, feelings of strangeness. It keeps couples going and able to deal with repeated failure. Interestingly, this is the same language used by clinicians to rationalize their own role as medics. Clinicians need to believe in the technologies too.

The Provision of Services in Private Clinics of Assisted Conception

The lack of a prescriptive legal framework and the free-market status of infertility treatment influences how the practice of gamete donation is presented, marketed and sold outside and inside clinics by clinicians

BOX 7
Note on the Lack of Statutory Law on Assisted Conception

At this point in time Italy is a free land for reproductive services, the so called 'Wild West of assisted conception' and a destination for reproductive tourism (see Appendix I). The legislative vacuum makes it possible to meet desires that could not be met elsewhere.[1] Only very minimal control is imposed, exclusively affecting the public sector, and leaving the private sector without a legislative framework.[2] The first administrative act, the Circolare Degan 55/1985[3] issued by Minister of Health Costante Degan in 1985, gives to married couples using their own gametes access to treatment in the public sector. This provision, regulating only one area of public infertility services, leaves all other treatments in the hands of the private sector; in particular all those treatment involving the use of donated gametes which are widely perceived as the most controversial. The private sector is left to run programmes of gamete donation and importantly the selection and screening of donors that such treatment require. The second administrative act, *Ordinanza Bindi*, issued in 1997 by Minister of Health Rosi Bindi, forbids the commercialization of the gametes themselves as well the payment to donors (it only allows the reimbursement of expenses). Special measures (*Ordinanze*) also prescribe the screening of donors for genetic disorders and impose the use of frozen sperm to prevent the spreading of the HIV virus. Further, at this time the national registry for assisted conception (RNIPMA – *Registro Nazionale Italiano di Procreazione Medico Assistita*) is instituted for centres and clinics of infertility to register on a *voluntary* basis.[4] Apart from the two administrative acts above mentioned and the special measures nothing else is legally enforceable. In 1995, the National Order of Clinicians (FNOMCeO – *Federazione Nazionale degli Ordini dei Medici Chirurghi e degli Odontoiatri*) makes a number of

---

1. The legislative vacuum also creates ample space for equally infamous cases of malpractice beyond the acceptable margins of health and safety regulations to became known nationally and internationally. In 1997 (whilst I am carrying out fieldwork) over thirty clinics are discovered not to be adequately screening donors and to be using hepatitis-infected sperm (see Owen 1997).
2. A survey of the *Instituto Superiore della Sanità* carried out in 2001 reports that at the time there are 384 centres for infertility treatment, both public and private (2003).
3. *Circolari* and *Ordinanze* are administrative acts issued by a Minister to regulate an area, or a specific aspect, when and if a law is absent. Degan is the surname of the Minister who issued the *Circolare*.
4. *Ordinanze* are temporary measures to intervene in specific areas in the absence of legislation. These are the most significant: Ordinanza Ministerale 5 marzo 1997 on *'Divieto di Commercializzazione e di Pubblicità di Gameti e di Embrioni Umani'*; Ordinanza Ministeriale 5 marzo 1997 on *'Divieto di Pratiche di Clonazione Umana o Animale'*; Ordinanza Ministeriale 4 giugno 1997 on *'Proroga dell'Efficacia dell'Ordinanza Ministeriale'* 5 Marzo 1997 *'Concernente il Divieto di Commercializzazionee di Pubblicità di Gameti e di Embrioni Umani'*; Ordinanza Ministeriale del 25 Luglio 2001 on *'Divieto di Importazione e Esportazione di Gamete e Embrioni Umani'* (see Di Piero and Casini 2002)

Conceiving Kinship

BOX 7: continued

recommendations for practice in the Code of Medical Deontology (*Codice di Deontologia Medica*). This is an entirely self-regulatory initiative which has only minimal impact on practising clinicians. It is a sort of warning against controversial practices such as surrogacy arrangements and treatment for lesbian and gay couples, in the absence of a legislative framework.

The Italian case constitutes an anomaly within the European panorama for at least three reasons. First of all, programmes of gamete donation – widely perceived as the most problematic and controversial by many – are left in the hands and monopoly of the private sector. Private clinics feel entitled to put in place their own procedures for the selection and screening of donors, the collection and storage of gametes, the number of donations per donor and so forth.[5] (During fieldwork I was asked several times if I could procure friends willing to donate to the clinic!) Secondly, the entire system of private provision is left to operate above the scrutiny of any authority. For many years there has been no audit – no *super-partes* authority entitled to check the overall standard of services provided by the clinics. Clinics are not registered on a compulsory basis, and clinical practice is not routinely inspected, unless a case of evident malpractice come to the attention of the authorities – the NAS, a special police body.[6] Thirdly, the private sector has been enjoying the benefits of a highly profitable market not because of a voted bill but because of a lack of it. Services and treatments are located in the marketplace and pushed into the arena of commercial services – entirely managed as businesses – due to the inability of the Italian legislative and political body to approve legislation.

---

5. It is worth noticing that often clinicians claim to be operating like their European counterparts; for instance, the Italian organization *Cecos* (which groups a number of private clinics) claims to follow the French system and to have 'borrowed' the French acronym (but apparently the French have disclaimed any connection long time ago (Novaes 1999, personal communication).

6. Elsewhere in Europe assisted conception, and especially treatment with donated gametes, have been regulated since the early 1990s (for a general overview of legislation in Europe at this point in time see Report on Bioethics of the European Parliament 1992; see also Gunning and English 1993; Nielsen 1996; Novaes 1986 for the French case; Morgan 1993 for the British case). Morgan, for instance, writes for the British case that the Human Fertilisation and Embryology Authority 'as [a] licensing body ... has power over public and private institutions to scrutinise and license, to approve, discipline and sanction the provision of assisted conception services and three main areas of activity; the storage of gametes and embryos, research on human embryos and *any infertility treatment which involves the use of either donated gametes* or embryos created outside the human body' (1993: 80, emphasis added). Price also points out the pressure that the British Government has been under 'to regulate work with embryos ... exacerbated by the complexities introduced by the use of donated gametes' (1995: 177).

and, consequently, how couples ask for help and come to choose and settle with the programmes.[7] Outside Italy, an interesting example of the way in which the system shapes provision is that of surrogacy in Great Britain (see HFEA guidelines) in contrast to various forms of commercial surrogacy in the United States (see Ragoné 1994 for American commercial surrogacy and particularly the role of commercial agencies). The free market defines the provision of treatment and services, emphasizing a specific rhetoric: that of *personal choice*. Stolcke has already noted this: 'biologists, geneticists and doctors argue that they are only responding to peoples' needs and demands ...' (1988: 8). Italian private clinics argue that it is choice and demand that justifies private provision, and not the lack of availability of such services in the public sector, despite the fact that private clinics are the only places where programmes of gamete donation can be accessed. As a matter of fact there is no choice.[8] The *Circolare Degan* forbids the public sector from offering programmes involving third parties' gametes, so if couples wish to proceed with such programmes, they must rely on private services. Private clinics have full monopoly in the system (see note on legislation at the time of fieldwork). This pushes the provision of such services entirely into the arena of consumption, with some significant implications. In private clinics, programmes of gamete donation are construed as 'the solution' for overcoming irreversible infertility: the diagnosis of infertility is reshaped and made momentarily flexible. Couples' infertility is transformed from being an irreversible deficiency (in the public sector) into a temporary deficiency that can be alleviated (in the private sector). Of course what changes is not the diagnosis, but its description. The mere option of a programme of gamete donation temporarily changes its status. The new status is fictional; it is the fiction of the private sector. In clinics of assisted conception couples suffering from impaired infertility are turned into customers and made to believe that they can buy fertility services – ultimately fertility as such. Strathern writes for the English case: 'those who seek assistance, we are told, are better thought of not as the disabled seeking alleviation or the sick seeking remedy – analogies that also come to mind – but as *customers seeking services*. The new technology, meanwhile, enables persons to achieve desires that they could not achieve unaided, *not without the*

---

7. As mentioned in the Introduction, and in line with previous chapters and more broadly with anthropological modes of writing, this chapter also is written in the ethnographic present, although programmes of gamete donation are at the moment no longer available in Italy.
8. The public sector – a hospital or a centre for infertility treatment – is the first point of reference for many couples with fertility problems. Couples are referred there (by their GP, *medico di base* or a specialist). Most couples begin treatment there, but move to the private sector, usually after several unsuccessful treatments with own gametes, if they decide to opt for a programme of gamete donation.

money to buy the techniques' (1990: 5, emphasis added). Miller also writes: 'in modern market and political rhetoric, services are increasingly also seen as pseudo commodities, where clients are turned into customers and the organisation of supply is supposed to model business practice' (1995: 11).

Stefania (married to Mattia):

When the doctor tells you that the sperm is not good and that you may try again but it may not happen you need to have another hope somewhere, and when the doctor [clinician who works in the public sector and in private clinic too] suggests to go private so that you may have more chance, you go. We came here [in the private clinic] a year ago. Doctors made me feel better and explained many couples went through this and finally had a baby. So when they suggested an *eterologa* I thought there were so many other couples doing it. I am very grateful. I feel this is now going to happen. *The doctor made me choose the right thing.* (emphasis added)

Francesca (married to Marco):

We came here because the doctor [clinicians met in public hospital who works in private clinic] told us this was the right place for us. He told us about egg donation. We then *choose* to go ahead. (emphasis added)

Michela (married to Daniele):

In a private clinic the doctors know best what to do [the clinician work in public hospital and in private clinic]. Here they can offer the best of treatment. You need to leave them to do their job. They know what it is all about much more than we do. When the doctor said to come and see him here [clinician met in public hospital] I knew this would mean that he would take care of it. In the hospital we were taking care of it. There was a lot more we had to think about and it did not work out.[9]

As soon as couples enter the private sector, they find themselves caught in multiple dynamics, and a new set of rhetoric about the 'public' and the 'private'. In private clinics clinicians operate *personally*, and play a visible role in shaping couples' choices and decisions. Besides being the immediate providers of information about infertility issues and the programmes themselves, they design the programmes. They are providers of infertility services. In the clinic, *clinicians sell the technologies.* They market them to couples, and couples will experience that in the programmes. Ragoné (1994) found the same with 'open' and 'closed' surrogacy programmes in the United States. Different programmes

---

9. This is a typical statement of how certain misconceptions about the public are forged in the private by clinicians. In the public hospital this couple were undergoing treatment with their own gametes (as treatment with their own gametes is the only available treatment there). In the private clinic they undergo a programme of gamete donation, making any comparison unreasonable.

played a fundamental role in forming the experience of couples and surrogates and this was evident not just to the observer but to those involved in them. The programme makes a difference as it shapes the experience. [10]

Life around Clinics and Clinicians: Trust, Faith and Dependency

The clinic is a physical space where conception becomes explicit and non-intimate. It is dislocated in the unfamiliar; the act of conception becomes a *mediated* enterprise requiring medical expertise. Particular alliances with clinicians come into existence (Price 1992), as much as complex relations of power. In the eyes of couples, clinicians are eventually able to produce the *unproducible*. They hold an exceptional expertise, with which they control the procreative project and possibly make it happen. Clinicians have great authority as they alone know the real state of affairs: they know more than anyone else the state of health of the couple, they make the diagnosis, and they provide the remedy. They provide (and choose) the gametes, and run the programme. Couples nurture *faith* and *trust* in them (as much as in the entire system of private treatment).[11] They put themselves fully in the hands of clinicians because they are their last hope. They firmly want to rest in those hands.

> Emma (married to Guido):
>
> The doctor told me to trust him [*avere fiducia*] and have faith [*avere fede*]. Once I was so depressed that I could not be examined, then he said to lie down and calm down, that he knew what he was doing and promised I would have a baby sooner or later. I felt a sense of relief. I often think that if he did not say that, in that moment, I would have left and never gone back. I was very tired at that point.

---

10. Couples who attend programmes with their own gametes, couples who adopt and couples without vested interest express strong reservations, in different ways and with different emphasis, about putting themselves in the hands of the private sector at a time of no legislation. They perceive programmes of gamete donation as particularly controversial and the free monopoly of the private sector as highly problematic, especially with reference to the screening, selection and record keeping of donors' information.
11. Although Italians use the private health sector for specialist consultations, generally speaking the relationship between the public and the private health sector is highly politicized. Face-value trust and faith should not be taken for granted at all.

Cinzia (married to Michelangelo):

Doctors can eventually make this happen. This is the thing: they can do what you cannot by yourself. I cannot but trust and hope in the miracles of modern medicine.[12]

Michela (married to Daniele):

Because of doing it[13] [the baby] with the doctors [*facendolo con i dottori*] I may have my own child and perhaps even a better one.[14]

'Faith' and 'trust' are used in private clinics as a mean to various ends, and are circulated widely. Rabinow (although in a different context) reminds us of Bourdieu's point that *faith* is a precondition of society. He writes: 'society could not exist without an ontologically rooted epistemological blindness. Given his premises, he [Bourdieu] is thoroughly consistent in arriving at an understanding in which humans are universally blind to deep meaning of their own acts, a stance Alan Pred calls "epistemological rooted ontological blindness"' (1996:11). A stance which is hard to sympathize with but that is particularly poignant in the context of programmes of gamete donation and that strongly resonates with what one observes in couple-clinicians relationships. Couples seem to accept unconditionally what they are told by clinicians and never raise any doubts. They never question any of the actions clinicians take: on the contrary, they scrupulously listen to the 'doctors', and have absolute trust and faith in their words and actions. They do seem caught in that sort of blindness just described.

Such attitudes generate high levels of *dependency* – a dependency on clinicians who may be able to offer 'the gift of a child' (*il dono di un bambino*) or 'turn the dream into reality' (*fare del sogno realtà*)', as both clinicians and couples use these expressions. It is a dependency that is almost tangible in the clinic: couples' lives revolve entirely around the clinic. When couples offer the details it emerges powerfully how they meticulously follow every indication that clinicians give, engaging and re-engaging as many times as they are advised to do so in examinations, tests and programmes.

---

12. This quote shows how faith and trust make the clinician appear (via modern medicine) almost able to do miracles. The clinician almost transcends his own expertise. Cinzia is having artificial insemination by donor, a very simple procedure. What she really has to hope for is to be provided with quality sperm to increase the chance of a pregnancy. There is no need for miracles at all.
13. A literal translation would be 'making it'.
14. This note reflects common assumptions about technologies versus health, as if making a baby in a clinic for assisted conception guarantees per se a 'better' ('healthier') baby. This is one amongst many assumptions, subtly perpetuated by clinicians: technologies of procreation could lead to 'healthier' (a questionable notion per se, see Rapp 2000) babies only if genetic screening was involved. Of course this is not the case here.

Letizia (married to Roberto):

I do what the doctor tells me to do. She knows and she wants to help us. I never object to an examination or anything else. I told her to do whatever is necessary to have the baby. I told her we do not look at the expenses.

Matilde (married to Giorgio):

You never know, you see. It is better to listen to the doctors here because after all they know what they are talking about. I make sure I follow what they tell me – the drugs to super-ovulate – you need to make sure to do the injections at the right time, otherwise it does not work. They explained it is a matter of minutes; otherwise it may not work anymore. With my husband we are very careful to listen to everything, to note down everything. You should see our kitchen – it is full of notes everywhere. We are also becoming a bit superstitious after all we have gone through. And this is just the beginning.

The state of mind is one of dependency. Couples (almost compulsively) hold in their minds images of what happened during a consultation, an examination and, of course, the treatment itself. Couples say that they find themselves discussing and rediscussing every single event: the way things went on one or other occasion, what the clinician said in this or that meeting, or the nurse or whoever else happened to be there. Couples report that they are always thinking about the clinic and the clinicians; they say they find it almost impossible to switch off.

Barbara (married to Giancarlo):

It works in a way that you are always here, even when you are somewhere else. Either because of practical matters or because it happens that you think about it – maybe you are doing something completely different and it comes to your mind or you watch a film and start rethinking this or that.

Giancarlo adds:

It is an experience that absorbs me completely.

In such a state, couples entertain continuous *imagined* relations with providers of treatment – imagined relations that they describe as being 'in the head [*è tutto nella testa*]'. The clinic and clinicians represent the world for couples, and everything else becomes ephemeral and collateral – in a sense, almost irrelevant.

Stefania (married to Mattia):

It all happens in your head, if you see what I mean – to go to the clinic, and see the doctors, is the most important thing for me now. It comes before anything else. I think all the time about this place, what happens here, what they say to me. I am sure they do not think about me that much!

Elisabetta (married to Germano):
I think that by now the doctor cares about me. She knows me well now, always understands my mood. Sometimes I feel like she is sorry that she cannot do more for me. I feel that even if she does not say it I come into her mind as she comes to mine.[15]

The Hyper-*medicalized* Infertile Couple

The hyper-medicalization of the couple is the most powerful representational strategy in infertility treatment and particularly programmes of gamete donation.[16] It sustains the practice. With it clinicians *normalize* any programme in Thompson's sense: 'normalization includes the means by which "new data" (new patients, new scientific knowledge, new staff members, new instruments, new administrative constraints) are incorporated into pre-existing procedures and objects of the clinic. ... The use of the concept of normalization incorporates both "normal" and "normative"' (2005: 80). If the couple is *ill*, clinicians can continually establish reasons for undergoing treatment – particularly when the programme fails. If the reproductive body does not work, if it requires therapeutic intervention, then the couple has a *medical reason* to attend the clinic. Novaes writes: 'the medicalization of AID thus appears

---

15. Overall, couples who attend programmes with their own gametes, and particularly couples who adopt, strongly feel the desire to regain their life – the former say that if the programme does not work they will take a pause and then think about what to do. The latter say that they have tried adoption after taking some time to reflect and disengage. There is thus investment in other aspects of life – possibly on life as a whole beyond the treatment. Paolo, who with his wife is attending treatment in a public hospital (therefore they are using their own gametes), says: 'The important thing is always to be reminded that there is a life out there and that this is not the end of the world. I always tell Elena that when we are done with the hospital: that's it – we need to think about something else – life must continue'.
16. This is quite a striking strategy as, of course, it is paradoxical if looked at in the light of the strategy mentioned at the start by which couples are made to believe that they are customers asking for services – services available in the marketplace – and making personal choices. In that case they are not made to think of themselves as patients looking for therapeutic aid. I came to make sense of such idiosyncrasy when I realized that it is possible for clinicians to put in place different strategies – and activate them according to the specific context – because of the very particular *clients* they are dealing with: these are couples who are very vulnerable since the desire to have a baby is so overwhelming (that's why they have to come to choose gamete donation) and therefore they are more prone than others to blindly accept multiple accounts without seeing certain internal contradictions and their instrumental use.

as an attempt to create a socially validating framework for this practice' (1985: 579). Couples do not attend programmes because of their desires, but because they have medical needs. It is remarkable how in a private clinic the desire for a child transforms itself into a disease; often, multiple diseases (see Mol 2000; Finkler 2000). By presenting the programme as a medical therapy, rather than as an intervention that gives access to the genetic material of third parties, a culturally upsetting reproductive option is displaced. Strathern writes: 'the new techniques of fertilisation do not remedy fertility as such, but childlessness. They enable a potential parent to have access to the fertility of others' (1990: 6). Giving it the appearance of a therapeutic intervention could not be more of an irony, particularly when the procedure is applied to the healthy, fertile, reproductive body (as in the case of sperm donation where the recipient woman, fertile herself, receives a donor's sperm because of her husband's infertility). As Finkler points out, 'medicalization changes peoples' perspective on reality, on their being, and on how they experience the world' (2000: 176).

> Margherita (married to Pier):
>
> Really it depends on the way you put it and on the way the doctor supports you. When you start, it is a strange thing: on the one side it is more difficult than you thought. When the doctor puts that syringe inside it is really weird. On the other side it is really a medical thing: you take drugs, come to the clinic, have scans, it is like when you are ill and need to sort out your illness; [infertility] is not really an illness, but it looks like one: also because of the situation, you are on the gynaecological bed, in a kind of surgery room, the doctor speaks to you in that way and everything is medical...you cannot *really* say you are making a baby [in this way] *it does not sound an extreme thing to do.* (emphasis added)

> Giancarlo (married to Barbara):
>
> I think the whole system is designed for you to feel as much as possible like a normal person, doing the most normal thing in the world. You buy the whole package like a package holiday.

Hyper-medicalization serves the purpose of obscuring what a programme of gamete donation ultimately does. The extreme emphasis on medical aspects and procedures helps to distance third parties. It turns the attention away from ideas of donors, acts of donation and bodily emissions. Hyper-medicalization gives couples (and clinicians) an altogether different focal point – a neutral ground to talk, and progress with the programme – re-engaging with it when and if it fails. As Gioia explains, the programme fades 'the worst bit of it' by medicalizing the intervention and emphasising the benefit of being '100 per cent under medical control'.

> Gioia (married to Leopoldo):
>
> It helps that things are organized in a way that, yes, on the one hand you get all that comes with it, which is really the worst bit [donation] but on the

other hand it is also a way to take care of yourself and your body in a medical way. I am going through lots of tests and exams, which is good because it helps to prevent other diseases. I am 100% under control and my husband too so it is good. It is like having a doctor in the family! If you understand what I mean.[17]

Ultimately, hyper-medicalization contextualizes gamete donation, making it acceptable. Culture is medicalized to reproduce culture. This is what makes it possible for all those involved in the day-to-day life of the clinic, clinicians and couples, to never ponder on the significance of perpetuating such manipulative devices. I hope I may be forgiven for calling it a *reciprocal farce*.[18]

Managing Recurrent Failure in the Clinic

In the daily routine of an infertility clinic the programmes, as pointed out earlier, are said to be therapeutic interventions, more or less equivalent to other medical treatments. A programme, just like a therapeutic enterprise, needs to be normalized and assimilated into medical routine. In everyday practice couples cannot be continually reminded that such medical interventions are experimental, risky and frequently unsuccessful. This is how clinician Viola explained it to me whilst scanning the ovaries of a woman:

---

17. Again, couples who attend programmes with their own gametes and have no intention of switching to programmes of gamete donation, and particularly couples who adopt, seem to have more reservations about being heavily medicalized. They often talk of 'doing too much to your body'. They look at technology with reluctance, as Paola, an adopting mother herself, puts it: 'technology interferes too much with nature'. Francesca, another adoptive mother, who previously failed two cycles of IVF with the couple own gametes, explains: 'It was quite heavy for me to go through IVF. I found the treatment unbearable and when we failed the first time it seemed insane to do *all* that to my body, to take that many drugs and to have the surgery done. I tried a second time for my husband – but I would never do it again. It was a relief to get out from the hospital, the doctors, the overall medical thing. The first thing I did was to go back to my homeopath – the second to start the adoption process'.
18. I wish to apologize to couples, of course, as I entirely empathize with their extreme pain and desire for a baby and have no right to assume that in similar circumstances I would have acted differently. I am of course speaking as an observer and as an external, with the distance thereby of someone who sees things from a non-emotional and non-implicated perspective. From this particular position I can say that in the clinic, and witnessing the relationship between clinicians and couples, one notices the performative aspect of it and the sense that there seems to be no pondering. Couples undergo one treatment after another, take drugs to super-ovulate, undergo surgery to extract the eggs and so forth – somehow one perceives no sense of limit and perspective.

> The NRTs are therapies which favour reproduction as much as contraception obstructs it. The NRTs are nothing more and nothing less than the result of certain applications in modern medicine. If you accept in principle human intervention on the human body, thus medical intervention on the ill body, you will have to agree with me that you cannot but accept fertility treatment as a therapy for the cure of infertility.

However, as soon as failure occurs the tale gets more complicated. The general perception of the technologies is not affected, *but only the experience of the couple who is encountering failure*. Suddenly, with respect to that specific treatment, delivered to that specific couple, that specific medical intervention is no longer talked of as a medical therapy. Contrasting notions are put forward to address different, opposite and simultaneous reasons depending on the case or moment or, sadly, the couple. Inconsistent as it may seem, failure is soon linked to the couple undergoing treatment. When failure begins to occur, the couple begins to be invested with responsibility – and guilt. The possibilities, chances and outcomes of medical intervention become plastic and the programmes become increasingly 'exceptional medical procedures'; exceptional, according to the clinicians, because of what they call 'the specific couple'. It is the couple that turns such therapeutic interventions into something that may work or may not. Not the other way round. Clinician Viola explains it again (in the course of a consultation with a couple who repeatedly failed):

> We explain to couples that these are complex technologies, you see, the latest advances in the science of reproduction; when we incur an unsuccessful treatment we explain that the same result cannot always be achieved.

A few minutes later, still during consultation (the couple is sitting opposite us) clinician Viola points out:

> This particular couple for instance has a negative response to the procedure. I have already discussed it with both of them, did I not? [Viola smiles to both wife and husband]. *It happens sometimes, not that often, but you see we are all different and unconsciously may not accept treatment.* (emphasis added)

Failure is never related to the specific medical practice and a clinician never takes responsibility for failure, nor does he/she suggests the possibility that, for instance (the examples and possibilities are numerous), the quality of donated gametes (including collection and conservation measures), the standard of procedures (manipulation of gametes and conservation of embryos), or technological expertise (surgical skills) might have played a fundamental role during treatment. The fact that *all* these aspects contribute to success or failure is consistently, and deliberately, disguised.

> Clinician Rosso:
>
> We *touch* extremely complex mechanisms; we enter into a space dominated by the laws of mother nature. *We above all deal with the personality of many*

different couples, their physical and psychological response. Often it is the patient who responds in an unpredictable way. (emphasis added)

Thus, in the words of the clinicians, unpredictability resides in the couple and not in the complexity of the medical intervention. It is for the couple – materialized in the person undertaking treatment (the woman) – to respond to medical intervention. As Becker has pointed out, 'the female body, in particular, has been reconceived as a site of technological advance' (2000: 6). It is up to the disciplined (female) body to respond (to paraphrase Thompson 2005). The couple (the woman) can respond well or sometimes exceptionally, but often does not. The couple, the woman, fails. As another clinician puts it, 'not everybody responds adequately, yet'. 'Yet', she says, perhaps implying that if the couple (the woman) were in the future to respond more efficiently, the technology would too. The more that medical intervention is advanced, the more that failure is encountered, the more the couple (the woman) is inept.

Getting to Understand Programmes of Gamete Donation

In a clinic of assisted conception there are two systems of communication in place: a system that exposes couples to a highly sophisticated medical and technological language – in Rowland's words, 'reprospeak' (1992); and a system that relies heavily on a language that I call the *language of commonplaces*. It is precisely the distance between the two that illuminates the artificiality of the device. Clinicians have a strong interest in using both, and the ability to use one or the other, at the right time and in the right circumstance, makes a clinician successful – that is, with a crowded waiting room and a long waiting list of clients willing to join the programme. The skill rests in the capacity to formalize or deformalize descriptive and explanatory medical procedures to suit each situation. It also rests in the ability of the clinician to shift easily from one to the other at the right moment. How are the two languages put to work? The formalized, hyper-medical language is used when the couple is challenging and wanting to gain more information and knowledge: the couple is for instance openly enquiring about treatment or is asking for specifics about recruitment, selection and screening of donors. The hyper-medical language is used to convey, and partly display, expertise – the clinician knows best, knows more than anyone else, the clinician should not be questioned, too many questions should not be asked. With the hyper-medical jargon an attempt is made to intimidate the couple – to keep the couple in their place. The more the couple is demanding, the more the clinician inflates the jargon to widen the distance. The deformalized language of commonplaces is used when the couple would still like to gain more information and knowledge but shows confusion, is not direct and finds it difficult to ask – and phrase –questions. A language of commonplaces is used to convey reassurance, and also partly to patronize the couple and the relationship. This is a language that usually triggers emotional responses. With the jargon of commonplaces the clinician normally shows empathy. S/he reinforces the relationship with the couple and leads it. Deformalization is introduced when emotional closeness becomes necessary.[19]

With both languages the final outcome is the same. The clinician is empowered against an increasingly disempowered couple – the longer the treatment lasts, the more the couple is disempowered. Contrary to what one may suspect, the use of one or the other language is not highly dependent on the class, professional or educational background of the couple (i.e. couples' ability to form and articulate questions, to be more or less inquisitive, is dependent on many other factors that influence their state of mind and emotional behaviour in a clinic). Social position may play a part in some obvious cases, but there are other dynamics that prevail and that bring clinicians to respond in one way or another. As mentioned earlier, a clinician is proportionally successful according to his/her ability to recognize at a very early stage what *touches* a particular couple. The clinician uses the language that serves her/his own interests best. Clinician Giallo (unwarily) offers a very clear explanation of the two systems of communication in place. He says:

> There are patients who have an *instinct* for medicine, they grasp what you are talking about. With them I can use the medical language and they will understand. There are others that need to be supported, explained what it is all about, with those you need to use different words. It is a bit like with my children, if I have to explain how the heart works, I explain it in very simple terms. I use the example of a pump that pushes air inside the wheel

---

19. I should make a point about the use of the language here. The use of a highly medicalized – technical –language is possible because, although couples come from all walks of life, they are very rarely medics. This is not a mere coincidence (none of the clinicians, embryologists, biologists interviewed ever said that they would themselves rely on programmes of gamete donation). Clinicians would often use a similar language with me when they wished to gain a position of power and obstruct my questioning. The lack of training as a medic makes it difficult, if not impossible, to endlessly sustain such conversations. Interestingly, in one of the clinics where I worked, such an attitude faded away when two clinicians realized that, although I was of course not a medic myself, I probably had sufficient familiarity with the terminology as the daughter of a clinician and head of a public hospital in Milan. From that moment onwards every question I asked was answered with another question that should have stimulated an emotional response. An example: when, at the beginning of fieldwork, I asked if donors were really matched with couples, the clinician gave me a sort of hyper-medical response of what he called 'the genetic matching'. When, later on, I asked him the same question he said: 'Let's leave out the field of genetics ... let me put it this way... if you were suffering from infertility would you want to be picky about the donor or would you want to make sure you become pregnant?'

of a bicycle: the heart works more or less in the same way, the heart is a pump. This is what I do with my patients. I make them understand what they need to know. But you see it is not really essential to know everything in life. It is also Okay not to know that much. (emphasis added)

I then ask how he explains to his patients (as he calls them), especially those who in his view do not have an 'instinct for the medical language', the genetic implications of using a donor, for instance. He replies:

> *I tell them not to worry, that children are always children, and it does not really matter the genetic component.* I tell them that we make sure we provide good donors, and we screen them and look for some similar physical characteristics. *I reassure them.* I tell them that I have two children and they are very different. You would doubt that they are brother and sister.[20] *I take it philosophically.* (emphasis added)

It is interesting that clinician Giallo, in explaining that there are different systems of communication in place, confirms that these are managed by clinicians on a case-by-case basis. It is interesting because when I asked if this was so, he replied with a definite no and was quite offended by my insinuation. I did not dare to add at the time that the communication system seemed also to make ample use of rhetorical devices. One can notice a similarly rhetorical use of language in the different context of the UK. Konrad (2005) highlights how clinicians in the context of non-IVF ova donation practices use certain discursive strategies, which overall seem to powerfully resonate in form and content.

The Work of Kinship in the Clinic

Clinicians heavily exploit certain notions to facilitate and propagate programmes of gamete donation. They use a specific terminology to evoke familiarity – in particular, they draw upon the domain of kinship. As Ragoné points out for the United States: 'Kinship terminology is routinely employed even prior to the conception or birth of the child' (1994: 39). The use of such specific terminology helps to direct, sustain and motivate couples at all stages of the process; it helps them to overcome doubts and fears, and to justify repeated attempts, effort and cost. A deliberate choice of a kinship language and of specific notions is made to reiterate very particular ideas about infertility as a negative and a deeply *unnatural* condition; the desire for children and so the desire to reproduce oneself as fundamentally *natural*; the family as the most desirable, if not compelling,

---

20. Of course! They are stepbrother and stepsister! Clinician Giallo always omits to say that he is divorced and remarried and that the two children were born from two different mothers. The majority of couples I talked to who were undergoing treatment with him would tell me the same story: 'The doctor explained that he has two children and they look so different anyway!'

achievement in one's life; and biological and social relatedness between kin as indisputably significant. These and a few other derivative notions are continually evoked in the clinic, in myriads of ways and forms, so as to keep them always present and in circulation. They are thus incessantly reiterated and heavily deployed with couples.

Clinician Marrone:

When a couple come to see me I explain what I can offer, and what I cannot. I immediately say that I cannot make miracles, but I also say that I may be able to give them what they are looking for: *a baby, a family, the joy of a birth. I explain to my couples that if we are lucky they may hold their baby soon.* With three or four cycles of artificial insemination by donor they can have a baby. In all those other cases in which the patient needs to undergo an IVF cycle the percentage changes, but it depends from case to case. I help them to believe that this is going to happen, a positive psychological approach is what they need. *They need to believe that they, as every body else, will become parents.* (emphasis added)

Clinician Marrone, as others, knows that this language, this particular kinship language, will calm couples who desperately look for a baby, and who incessantly attend the clinic and engage in one programme after another. He also knows that to reiterate an ability to provide 'a baby, a family, the joy of a birth' is, in those specific terms, precisely what couples wish and need to hear, to be able to sustain the strain of a programme and to re-engage. Most of all, this is what they need to hear in order to come to terms with the unsettling idea of donation itself, which is always in the background. For this reason, clinicians frame notions of donation relying heavily on the realm of kinship: donation is described as a safe practice for *family life*, the donation of gametes as an act *without moral and familial implications* and a solution which will, sooner or later, bring happiness *amongst all kin*.

Clinician Nero:

Couples need to understand that gamete donation is Okay, that *it does not really create complications in family life*. When the baby comes couples forget about the donation, I know this by experience. I always tell my couples about the experience of other couples, who went through this before. There is nothing better than first-hand experience. I want to make them understand that we have good statistics; *when the baby arrives, when the whole family sees the baby, they will forget.* (emphasis added)

Clinician Arancione:

I have always a preliminary meeting with the couple, to make sure that they are Okay, that there is no problem with the idea of donation. I explain that it may all look different because they come here, in the clinic, to have their baby, but it is really not that different, *the donation of gametes does not change their being parents.* In a sense it is a technicality. (emphasis added)

Like couples, clinicians empty donated gametes of their intrinsic properties and uniqueness. They also need to put the emphasis elsewhere: they resort to a complex strategy as they need to attribute kinship significance to the genetic tie between the fertile partner and the baby to be born (as otherwise there would be no reason to undergo a programme of gamete donation) whilst at the same time denying significance to the genetic tie with third parties (providers of genetic material). The link with third parties needs to be ignored and replaced with the social and emotional tie that indeed will be immediately established between the other parent (the infertile parent) and the baby.

Clinician Blu explains how he suggests to couples the least problematic way to think about gamete donation, gametes and the resulting offspring. He says:

> I am not only a doctor here, but also a kind of psychologist: *I always explain to couples that what makes a person is the social environment. It is important to be good parents, to be good mothers and good fathers.* Genetic inheritance plays a minimal role. I always point out to them the case of two twins who are separated at birth and live with different parents. They will be different. *It is the family that matters, the parents and the relatives.* (emphasis added)

At a different point in the interview clinician Blu also points out:

> I also always make clear to my patients that it is very important, of primary importance I would say, *to preserve what's already there in the family. What's already there is their treasure and they have to treat it as their family treasure* – with extreme respect, almost devotion I would say, particularly as that's all they have. (emphasis added)

It is because of the language of kinship, the ambiguous use of that language, that couples are sustained and sustain themselves throughout the various stages of a programme. The language of kinship is used to naturalize and normalize programmes of gamete donation. The language is special not in content, but in its capacity to evoke associations. As Bourdieu (1992) suggests, the exchange of words is an exchange in meanings. It is a language of very *selected* notions easy to reproduce. It is a language that reproduces known patterns of questions and answers. It is very predictable. It is this predictability that makes it an excellent medium in clinics of assisted conception.

CHAPTER 6

LESBIAN AND GAY COUPLES MAKING FAMILIES BY DONATION

A Case: A Lesbian Couple Planning a Family by Donation

Nora (in a relationship with Cristiana):
It happened between us, but it could have happened with anybody. We have been together since '92. I fell in love straight away. For Cristiana it was the first time. Until two years ago we did not even think about children. We started to discuss it when two couples, very close friends, had two babies at a distance of two months from each other. It was pretty shocking. From having no *nephews* to having two almost in one go. We always took it completely for granted that none of us would ever have children. At first it sounded quite odd –especially when they told us how they worked out the whole thing [they arranged conception respectively with two gay couples]. We discussed it a lot. *We wondered if it was really a desire between women or an unconscious attempt to recreate the lost family.* (emphasis added)

Nora's partner, Cristiana:
Initially, I found it problematic: it seemed to me such a difficult thing to do, to arrange and to agree with someone you barely know. I immediately thought about the children and the fathers – from the point of view of the children. It seemed quite a big thing to do, an extreme choice. I think I was the one, amongst all of us, more concerned and uncomfortable with it. Nora was much more positive, she immediately could imagine it. For me it is all much more recent. During a holiday in India, I worked out the whole thing in a new way. When *we returned we spread the word around [mettere in giro la voce].*

Nora and Cristiana illustrate the predicament of many same-sex couples who go through the process of planning families. They illustrate the stages couples often go through before deciding to act on the desire and make clear the complexity of such a choice, with so much at stake. After coming back from a holiday in India Nora and Cristiana have been contemplating more than ever the possibility of starting a family together, and have been proactive in making it happen. To spread the word around, *mettere la voce in giro*, means to let friends and acquaintances know one's intention to find the right contributor for conception: he can be an *unknown/known* donor or what I have called *a partner in asexual conception* (see below). When I met them, Cristiana and Nora were choosing between an unknown donor (a friend of a common friend, neither Nora or Cristiana knew him personally) and a couple of gay friends who seemed willing to have a child and share some form of co-parenting. Forms of co-parenting require complex negotiation and arrangements.

Introduction

In Chapter 3 and 4 I have presented the ethnography of heterosexual couples suffering from irreversible infertility and choosing programmes of gamete donation to conceive their *own baby*. There I have suggested a trajectory, as I have come to know it from couples: from the time prior to discovering infertility – a time of expectations and dreams – to the time of choosing a programme and treatment. In the subsequent chapter I have described the more arduous time that follows, when couples must reconcile themselves with the practice of donation as a whole. In Chapter 5 I have added the further dimension of the relationship between couples and providers of treatment, at a specific time when the lack of legislation creates an environment in which clinics are effectively unaccountable. This chapter jumps into a different world altogether. It explores the different ways in which lesbian and gay couples making families think about procreation, the family, the making of relatives, and biological and social relatedness precisely *as a result of* their different positioning in the world. It explores how these couples produce a different narrative and make sense of the ties between biological parents, co-parents, unknown/known donors and partners in asexual conception. As its aim is to illuminate by comparison the ethnography of heterosexual infertile couples attending programmes of gamete donation and the use of certain kinship constructs, it does not attempt to give a complete account of lesbian and gay contemporary lifestyle and family making. The project was never designed to be as wide.[1]

---

1. This chapter covers aspects that link with narratives presented in Chapter 3 and 4. Whilst it is illuminating in the context of this work, it has limitations in other respects. First of all, I will not be able to draw substantial distinctions between lesbian couples and gay couples and instead will often treat them as if they act similarly and are part of the same social world. This is not entirely the case and there are several differences that would be addressed if this work were to be focused on lesbian and gay couples alone. Such difference would constitute an analytical point. Having said that, I have to immediately contradict myself and point out that in the context of lesbian and gay couples making families there is a surprising closeness which is indeed absent in the wider lesbian and gay community (see Bonaccorso 1994). There the relationship is much more oppositional.

Lesbian and Gay Couples: Planning a Life Together

The lesbian and gay couples interviewed, like the heterosexual couples, have very heterogeneous backgrounds. Education, class, wealth, and profession vary and portray the general differentiation, once again, found in any urban milieu. They originate from, or live in, different parts of Italy. Political orientation also varies considerably, although the majority of couples claim to be centre-left wing, and only a minority to be right wing. No one claims to be a practising Catholic, although some define themselves loosely as believers. These couples are all in stable relationships, have lived together for at least two years, with an average of three to five years; few couples have been cohabiting for more than five years. Apart from very few exceptions they are all between twenty-five and forty-five years old, with the majority beyond their mid-thirties (see Barbagli and Asher 2001 for trends on lesbian and gay couples cohabiting in Italy). These couples are all in the process of making families or have recently done so. In previous research I carried out in the early 1990s most lesbian and gay couples planning families had been married and often had children from heterosexual unions. They were at the beginning of the process of planning families within same-sex unions (Bonaccorso 1994).

Lesbian and gay couples describe the union as the result of 'falling in love' with someone of their own sex. They emphasize the moment of meeting the other – the emotional-romantic and sexual encounter – and often challenge the definitiveness of certain assumptions about heterosexuality and homosexuality as fixed emotional-romantic and sexual states/desires (see Butler's call on the restrictive meanings of hetero/homo/bi sexuality 1999: 40–41; but see also Calhoun 2003 for a discussion of lesbian love and feminism). They seem to be increasingly moving from a politics of aetiology (*being born* lesbian or gay) and fixed identity (*to be* lesbian or gay) to a politics of *flexible positioning and mutable identities*.[2]

> Lorena (in a relationship with Manuela):
>
> If you asked me five years ago I would probably say that it would be unlikely to happen, but I would not have denied the possibility either. It seems to me that anything can happen to anybody any time in their life. At the time it would have just sounded like a remote possibility. It's so much a matter of chance. I met Manuela and I fell in love. If I had not met her perhaps I would never have experienced to be with a woman. I do not think I had a particular desire for women before, I cannot say that my lesbianism was something already there, repressed or what. I think we should all look inside ourselves and really wonder if it does exist this thing about being

---

2. It should be noted that such flexibility is a one-way trajectory from heterosexuality to lesbianism and gayism –the reverse process is really never envisaged. This becomes clear in other conversations where heteronormativity is seen as the alternative against which couples project their own difference. The normative is described as containing expectations of the kind well described by infertile couples in heterosexual relationships.

heterosexual, or gay, or even bisexual. To me it is more about what I feel and how I feel really.

Lorena's partner, Manuela:

We met and it happened. It was sudden and the most beautiful thing that ever happened to me in my life. If you asked me five years ago, I would have told you 'you must be crazy'. I did not realize at the time that you can actually move and change so radically. The beauty, and I think we are both aware of it, is that it could change again … It's beautiful because it keeps the relationship alive. You need to work very hard for it. But you also know that it is the most special thing you have.

A politics of flexibility strongly emerges also when, as many feel, lesbianism and gayism are fundamentally a way of life. Maura (in a long-term relationship with Silvia), who never experienced heterosexual relations in her life, and for whom to envisage a change in emotional and sexual orientation appears an improbable and remote prospect, put things quite similarly:

I never had a relationship with a man – to like and love women is the way it is for me. I have been having relationships with women since I was an adolescent. It is quite hard to imagine the other way round, even if I can theoretically contemplate the possibility that it could happen. Yes in principle it can happen, but it is highly unlikely, very unlikely actually.

Micaela (in a relationship with Lucia):

It's more about how you see things than what you actually do. In the late '70s it was all about separatism and the hard line. It was also about making so explicit that difference between us and them [heterosexuals]. Now, I think it is much more realistic. It is more about love and care, I think. In that sense, in principle anything can happen, if then it doesn't – for instance you keep falling in love with women, it is another matter. The politics is very different.

Lesbian and gay couples plan a life together choosing each other – for as long as it lasts, perhaps forever; as Weston says: 'Forever can, indeed, be a long time' (1992: 19). The relationship between same-sex couples is one of commitment, but also of freedom. Couples claim that the continuous re-invention of the relationship, the way to be together, to plan a life day by day and to conceive a future is the negation of the normative. They claim a lack of constraints, roles and rules. They say that in same-sex partnerships there are not a priori models (see also Weeks et al. 2001).

Sebastiano (in a relationship with Giulio):

The relationship is different because it is equal. You are never supposed to do this or that, there are no roles and rules apart from those you make. Day by day you literally create and reinvent your own relationship.

Andrea (in a relationship with Enrico):

It all comes down to your state of mind. With Enrico I am certainly not going to mimic my mother and father's marriage. As your brain automatically excludes that, it is always looking for something to use as a framework. But there isn't. How do you make a gay life? That's not something you learn. There is no imprinting. So you keep working it out yourself as a gay couple. It comes down to very small issues, like what you do in this or that case. Who does this and who does that. You try to keep it very fair.

Couples emphasize the lack of a normative space and, instead, continuous fresh coalitions. They see themselves as continually reshaping the relationship, their emotional and sexual predicaments. They see themselves as *imagining* their own life, making it day by day as they want it without having to follow any convention, or expectation, since there are no conventions or expectations to follow. This is a fundamental difference from conversations with heterosexual couples, and indeed couples who attend programmes of gamete donation, who on the contrary look for the norm and claim to wish to conform to it.[3] The way in which lesbian and gay couples claim imaginative and inventive ways to stay together, love each other, and plan the future, in the absence of models to follow and be compelled by – including same-sex stereotypes – is absolutely poles apart. In some ways, it may be rhetorical but it is absolutely divergent from the heterosexual narrative, so divergent that Silvia, for a moment, almost wishes that sometimes she had a model to follow because to reinvent one's own life and relationships everyday is truly hard work:

Silvia (in a relationship with Maura):

With Maura I feel free because I am free, even though this is a stable and perhaps forever-relation. I never feel or felt constrained by anything. Quite the opposite, at times, I wish there was a sort of example to follow. *Sometimes it is even hard to reinvent everything.* (emphasis added)

Gigi (in a relationship with Carlo):

There are so many times that you just have to face the fact that you haven't a clue about what's going to happen, and how you are going to make it work. But other times, it looks like that is more of a mixing thing. You put very different kind of ingredients to make your cake. Do you understand me?

The way same-sex couples frame their life and future points to the lack of expectedness and ordinariness. They wish to convey *a fundamental distance* from anything that stands for and represents the conventional.

---

3. Of course, the negation of the normative does not preclude lesbian and gay couples from predictable life cycles as indicated, for instance, by Slater for the lesbian couple (see 1995). It only excludes them from the repetition of certain normative patterns often found in certain heterosexual lifestyles.

Even when not mentioned, the conventional is heteronormativity. The heterosexual norm operates there, sometimes at the forefront, sometimes in the background, always as point of reference, as a continuous term of comparison. Lesbian and gay couples, claiming flexibility and the creative work that goes into the relationship, say they have an open outlook on the world: literally on everything. They say that very little can be taken for granted, above all the idea of *making a family*.

Planning Families

Lesbian and gay couples say there are different ways to make families. Whilst they see the project of the heterosexual family as, potentially, a heavily gendered and normative one – they ascribe gender-role constraints to heterosexual partnerships, whereby wife and husband fully accomplish their gender roles (see Connell 1987) – they perceive theirs as creative. They describe it as *imaginative*. They say that *overt* gender-and-power-role constraints are absent, because of the couple's composition but also because of the deliberate effort that they make. They point out that no one knows how to make a lesbian and gay family. They mention either freedom and creativity in thinking about families by donation or else a lack of assumptions about what it should be like. They claim to be investigating and exploring their own choices in a far more thorough way than anyone else possibly could: families are not pre-given, they are questioned and fully chosen (see Lewin 1993; Weston 1997; see also Weeks et al. 2001). Families are exploratory projects against the dominant model of heterosexual unions; as Saffron points out, they are 'against the grain' (1994: 5). In making families by donation couples are not trying to be assimilationist.

> Francesca (in a relationship with Susanna):
>
> We perhaps live on a different planet. I feel different from most people. I am quite proud of it. You cannot take it for granted that this [family] is going to happen, or that you really want this in your life. We do question if it makes sense to label our relationship; if it makes sense to be parents in this world and if it makes sense to add a new dimension to all we already have.
>
> Enrico (in a relationship with Andrea):
>
> This is about us making something that is extraordinary really. It is never going to be like them [heterosexuals]. We have been thinking about this for a long time now – questioning ourselves but also realizing that this is an entirely different project comparing it to everyone else's. It is hard for us as well to make sense of it! We wonder what it means exactly for us as gay men, we will know it only when this is going to happen for real – this is a phase of making sure that we are not locking ourselves into a trap [heterosexual]. Apart from that, we are going to make something exceptional.

Couples talk about the way they *imagine it to be* at the point in time in which they initiate their family project. They talk about difference and novelty, about non-conformity, a process which is indirectly encouraged by the official non-existence of lesbian and gay families, the general denial of lesbian and gay families (however stable and long term they are), the legal non-status itself. This all stimulates ample explorations of common notions: What makes a family? What is in a family? Why a family? In other words, the more lesbian and gay couples planning families are made to feel that they do not exist, or that they should not exist, the more they are impelled to ask themselves who they are, what they are, what avenues they should pursue. Questioning becomes an inevitable course of action: couples feel compelled to look for new definitions, embedded in new meanings.[4]

Francesca's partner, Susanna:

We both started playing with words: What are we? Who are we? A couple, women living together, lovers, friends having sex, lesbians. A family, what kind of family? When you enter into that, it never ends. You must go through each single word and work it out. At the end you always go back, more or less, to the words you started with. You understand that certain words such as being family, or being married, or being lovers *mean what you mean – what they mean to you*. I think that Francesca says we live on a different planet because of our playing around with everything, upsetting, whatever. Once you do it, even if you go back to use the word you started with, it is different because that specific word is now full of your own meaning. (emphasis added)

Lucia (in a relationship with Micaela):

We call it family because we do not have another word, but then it does mean something else. It's not a family like any other family in a normal sense. It's our family – a very special kind of family

As making a family, and having a baby, is not considered as *natural, common* or *usual*, same-sex couples find themselves revisiting basic assumptions (see earlier works such as Pies 1985; Polikoff 1987; Bozett 1987; more recently Lewin 1993; Benkov 1994; Saffron 1994; see also Ali 1996). These are often the very assumptions that inform the wider cultural milieu of families of origin, relatives, colleagues, and neighbours;

---

4. The lack of open recognition of lesbian and gay families is not just an Italian phenomenon. Weeks et al. (2001) amongst others treat the argument in their work. Civil partnerships came into force in the UK only on 5 December 2005. Civil partnerships now enable same-sex couples to obtain legal recognition. Amongst duties and rights is included the duty to provide reasonable maintenance for the civil partner and any children of the family, as well as the ability to apply for parental responsibility for the civil partner's child. The American literature (non-anthropological) is quite substantial; it goes back to the early 1980s and reports many of the difficulties encountered by couples and families in being recognized and accepted, as well as legal cases to gain custody.

assumptions that make it, at times, hard to share everyday life with outsiders, beyond the partner and the closest circle of lesbian and gay friends/kin by inclusion.[5] Lesbian and gay couples begin an exploratory journey, which exposes them to the consequences of persistent cultural investigations, mostly carried out with partners and a small group of friends. These are investigations with significant emotional and psychological repercussions: as much as they are sought after, they are also unsettling as they jeopardize the *cultural self* (see Cohen 1994). Couples often talk in terms of the cost of difference, of experimenting with alternative forms of living and of living a dissonant life. Giulio says: 'it is like sliding into the hell of Dante'; an uncomfortable image: once one is in it, it is hard to get out.

Stefania (in a relationship with Caterina):

We have all been thinking about this. We have gone through several critical periods in our lives: some of us just because of lesbianism, but not just that anymore. Twenty years ago that was more the issue. It was the main issue. Now it is more about how to make it our own way [the family with children]. This is perhaps what pushes many of us towards psychoanalysis. I reached a point where I was rejecting everything and did not feel I had any *real* alternative. It was devastating. (emphasis added)

Stefania's partner, Caterina:

We went through a very difficult period. It seemed impossible to reconcile our desires: wanting children, making a family without losing our sense of *exclusive women relations*. We did not want to imitate or reproduce petite-bourgeois families [*famiglie piccolo-borghesi*] but it was not that easy to create altogether something that different. We were also unsure about wanting to recreate something that was different just for the sake of it.

Giulio's partner, Sebastiano:

It's hard work. You feel that you have lost all your certainties. You actually feel that you are all over the place. It's like you are leaving a safe harbour

---

5. With 'kin by inclusion' I am not indicating that these are less kin-like; I only intend to distinguish biological/legal forms of kin relations from relations created by *selection* and *election* (see below). Lesbian and gay couples often speak about their social network as 'inclusive' and 'transformative'. This parallels Weston's American 'chosen families' (1997) and Hayden's notion of 'lateral families'. Hayden writes: 'Choice becomes the distinctive feature of these families, as gay men and lesbians consider themselves kin to people with whom they have no biological relationship' (1992: 3). It should be added that the attempt and desire to incorporate non-biological kin and out-laws into the kin arena is an increasingly common phenomenon in Italy: heterosexual singles and couples in their thirties and forties, often not-married and/or cohabiting, with or without children, frequently talk of others – emotionally close others – as sisters and brothers, aunts and uncles, to their respective children. Kin ties seem to multiply easily, far beyond the blood kin/legal realm.

and you have no clue of where to go, and what you are going to find out there. It's scary. Yes, it's quite scary. That's why we all end up in psychoanalysis. Do we know anyone who hasn't seen a shrink in the last five years? You move from being a gay man, and a couple, to a family with children. It's a completely different space.

Rethinking Motherhood and Fatherhood

Motherhood and fatherhood require even more work. They are put under scrutiny because of a pressure to reject assimilation to heteronormativity. They are seen more than anything else as potentially limiting and constraining if left unexplored. To borrow Delaney and Yanagisako's (1995) notion of *naturalization*, motherhood and fatherhood are rejected as naturalized desires. In particular, lesbian women question what is often culturally and socially represented as preordained – *essentially natural, normal, a desire unconditionally shared by all women*. They claim that common assumptions such as that women are born with a call for motherhood need to be fully revisited. It [the desire] cannot just be taken for granted on the basis of certain unconscious, induced, mechanisms', Sonia explains, 'such as the biological need to reproduce, so called maternal instinct, everlasting devotion and a sense of sacrifice'. They feel that the desire to have a child can be constraining if it embodies unquestioned notions of maternal fulfilment. They feel the need to rethink it as a *liberated choice*; a choice that a priori opposes constriction, limitation as well as any form of assimilation.

Paola (in a relationship with Elena):
If you want to have a child, you need to ask yourself why, where that desire comes from. You cannot just assume that it is OK. As women we are so much expected to want and long for a child that as soon as we feel that desire, we automatically believe that that is OK. But it is not necessarily that OK. As women we must learn to ask ourselves questions, and wonder *what we feel and want and why we feel that way.* (emphasis added)

Paola's partner, Elena:
Since I have been with Paola I have realized how much we all take for granted our own motherhood. In a lesbian relationship you cannot just say 'my dear, I want a baby' and make it. As soon as you say 'I want it' you also need to ask yourself, 'why do I want it?', That is just the beginning: you then start to ask yourself lots of questions about this and that, what it means to you, in your own life. What it means being a mother nowadays, what it means to be a lesbian mother in this world and so forth.

Manuela (in a relationship with Lorena):
There is so much that needs to be worked out. We both feel that we are just at the beginning. The big question is what kind of mothers are lesbian

mothers? These are the sorts of questions that you need to ask to make sure that you are not just taking it too much for granted.

Gay men also go through a process of revision, although they are confronted with different questions altogether. Fatherhood is historically less constraining. Men do not experience equal cultural expectations of paternal fulfilment. On the contrary, they often claim to have to reassert their wish and even their right to fatherhood (see Bozett 1987; Barret and Robinson 2000). This inevitably marks in a different way the process of rethinking fatherhood.

Maurizio (in a relationship with Bruno):

I grew up as a gay man thinking of not becoming a father. I went through a long difficult period: I had to convince myself, allow myself to think 'why not?', 'what's so wrong with being a gay father?' As a gay man you put yourself automatically into the category of singles. Then I ordered an American book about gay fathers – someone called Bozett I think (I do not really speak good English, I had to read the entire book with the dictionary!) but I managed to get through – in the end I thought I could make it if I met the right partner. And I think I have now. It has taken me more than ten years or so to get to this point.

Claudio (in a relationship with Marco):

My problem has been more about asserting my need for fatherhood [*il mio bisogno di paternità*] after such a long struggle to be openly who I am. I am a gay man and want to remain so. I want a child without having to play the role of a father like every other father. I do not fit into the heterosexual-like father, nor in that of the *padre checca* [the camp father]. I am going to be myself as you see me now. It's pretty clear that my family [of origin] expect me to behave as if I were a normal father, a father like everyone else, perhaps a bit more rigid, almost like a *padre-padrone*, perhaps to compensate for my being *on the other side*. This is the problem, as soon as you are different, no matter why, all the rhetoric comes into it. They expect that you do things that they would never even dream you did under different circumstances, that in those circumstances they would have found completely unacceptable.

Giulio (in a relationship with Sebastiano):

I came out at work six months ago, and since I have said that sooner or later we will have a child, colleagues who are generally surprisingly supportive (there are only a few who are deliberately awful and nasty) keep asking all sorts of questions. It's quite challenging sometimes, but you know, it does make you realize that there is really nothing you can take for granted about this. It is much more difficult for us as gay men, I think, because everyone seems to expect that we do not have a desire for children, to settle down, make a family and so forth. People are most inclined to think that all we want is to sleep around, change partners and have that sort of life. The kind of questions that people keep asking make you think quite a lot about you as a father, what you are really looking for, and what it means to you, how you are going to become a father and what kind of father you are going to be, how you are going to sort out the mess that at the end of the day you are going to create and so forth.

There is more. The need to explore where the desire for motherhood and fatherhood comes from is particularly important because of the particular historical moment in which lesbian and gay couples are putting in place their maternity/paternity projects. As it emerged from conversations with lesbian and gay Italian activists at the time of fieldwork, the (Italian) gay and lesbian movement (ArciGay/ArciLesbica), in its organized form often considers the desire for parenthood as nothing else but *expression of heterosexuality*, apart from a few exceptions.[6] Amongst others, one exponent of this position is gay writer and journalist Giovanni Dall'Orto (1998, personal communication).[7] The non-parenthood option amongst lesbians and gays who decide not to plan families and who are active in the wider community is emphatically voiced as the *real* lesbian and gay option. Any other alternative is rejected in absolutist terms. *Rejection is made as a statement of distinction.* Evidence that, *elsewhere*, same-sex couples look for and achieve parenthood without compromising on their lesbian and gay way of life is often strongly disputed. The American lesbian baby boom of the late 1970s, mostly developed in the mid-1980s on the West Coast, and the following gay boom of the late 1980s is perceived as assimilationist. When mentioned, accounts such as Weston's on lesbian and gay chosen families (1997) as well as Hayden's of American lesbian women making procreative and family choices (1995; see also Lewin 1993) appear as destabilizing ventures. The position of the wider community is quite rigid, but pervasive all the same. It puts same-sex couples opting for motherhood and fatherhood in a conflictual position as individuals, as couples and more broadly as lesbian and gay social actors. Couples wishing to become parents fight a cultural and social war not only (as it might be expected) against the heterosexual realm, but also often against the adverse lesbian and gay one. In order to act on their desires, they often have to go through a *cleansing* and *distancing* process.[8]

---

6. During fieldwork sporadic cases towards a politics of acceptance took place. For instance, in 1998 the president of the lesbian Movement ArciLesbica initiated a campaign to open centres for self-insemination and distributed a kit to practise it but it was not particularly successful or of great resonance.
7. I must also stress, though, that Dall'Orto's support in recruiting couples through the magazine *Babilonia* (a national publication on lesbian and gay culture) has been invaluable throughout my fieldwork. At the time of revising this work for publication his position, as that of many other lesbian and gay opinion-makers in Italy, has shifted towards acceptance. Even within the official lesbian and gay movement there are known cases of parents with children. For a general overview of the Italian gay movement see Rossi Barilli (1999).
8. At the time of revising this chapter the situation seems generally much improved compared to the time of fieldwork and to previous years, although there is a long way to go. My work on lesbian and gay parents

Carolina (in a relationship with Tania):

I do not know how many years it took before I was able to accept and reconcile my attraction for women with my desire for motherhood. This was the first step. After, I went through a period of practical thinking. My main concern was how to make my project happen without getting lost *in my own world*. (emphasis added)

Bruno (in a relationship with Maurizio):

There was a period that I rejected whatever felt hetero, to become a father seemed undesirable. It has been only recently that I realized how crazy I was. *We all were* [a group of friends] *obsessed with being gay and different.* I went into therapy because I started to have *contrasting desires: I liked men, but also wanted to have a child and be a father.* Now I am finally ready for it, and it does not feel that *hetero* at all. (emphasis added)

Lesbian and gay couples thinking about motherhood and fatherhood strategically distance themselves from the official lesbian and gay Movement, which frequently ostracizes their choices. They dispute the notion of inconsistency between lesbianism/gayism and parenthood. In the absence of a politics of acceptance, they relocate themselves elsewhere, in a space that once again can be rendered with Anderson's notion of an *imagined community* (1991) of others like them. This space is not necessarily in between the heterosexual and the lesbian and gay realm. It is possibly more of a new *third* space, which can accept another way to be, to think and to act. Hayden hints at something similar when she says: 'Lesbian co-parenting families *carve out a distinct place* in the tension between straight, biological families and gay, chosen families' (1992: 3, emphasis added).

(Bonaccorso 1994) was the first one on the subject in Italy and for long time there has been nothing else. More recently other works have been published which treat the argument of lesbian and gay families, such as Barbagli and Asher (2001), Bottino and Danna (2004), Borghi and Taurino (2005), Paterlini (2004) and a very informative study of the Gruppo Soggettivita' Lesbica (2005). In addition, for the first time, in 2004, a few gay families have come out publicly forming the *3G Group*, and in 2005 the association *Famiglie Arcobaleno* was formed. The latter is an association of lesbian families who have had children and continue having children either via assisted conception (now going abroad, as the 40/2004 legislation on assisted conception makes it impossible for them to access any treatment in clinics) or via what I have called 'informal networks' as described later in this chapter. This is an absolute novelty in the Italian scenario; it was absent at the time of fieldwork and indeed in my previous research carried out in the early 1990s. The first ever conference on gay and lesbian families, *Famiglie Omogenitoriali*, took place in Milan recently, creating much rumour. I am under the firm impression that we are witnessing the beginning of a *coming out* process – how this reflects the real heart of the phenomenon is a different matter. It may take years before we gain a full picture.

The Lesbian and Gay Way: The Procreative Project

Lesbian and gay couples plan the procreative project in various ways: some through (hetero)sexual intercourse with just the aim of procreating, some in clinics of assisted conception to receive artificial insemination by donor (AID);[9] and the majority outside the clinical arena.[10] Here I focus on the latter as they represent a new phenomenon and a more interesting one. These couples arrange conception using the language of assisted conception (donors and donations) and what one may call low-tech applications. Precisely because the issue is not infertility per se, conception is mostly achieved with self-insemination (SI). These are couples who make plans either with an opposite-sex couple or partner who may be lesbian or gay, or more rarely heterosexual. When the process is initiated by a lesbian couple (I call this *initiating couple*) three main options are available. The provider of biological material can be *anonymous* to the couple, thus an *unknown donor* (*donatore sconosciuto or ignoto*). This type of arrangement does not fundamentally differ from that experienced by infertile heterosexual couples suffering from impaired infertility and attending programmes of gamete donation in clinics of assisted conception. Or he can be a person met through a mutual friend with the explicit aim of conception – this is quite a common arrangement and the person in question acts as a *known donor*; he is chosen by the couple and he often chooses the couple too. He can be someone that was previously unknown to the couple or he can be someone who is introduced to the couple with the clear intention to achieve conception. In both cases of unknown and known donors *(donatore conosciuto or noto)* there is often a go-between *(intermediario)* who helps in the process, unless the donor is a close friend of the couple and existed in the couple's life prior to the

---

9. During fieldwork, clinics of assisted of conception were relatively free to treat lesbian couples, although in 1995 the *Consiglio Nazionale della Federazione dei Medici* (National Council of Clinicians) approved a self-regulation recommending clinicians not to offer treatment to same-sex couples. I have interviewed lesbian couples claiming to have attended clinics of assisted conception. I have also met some couples directly through clinics. Apparently clinics of assisted conception also used to treat both lesbian and gay couples together, keen to medicalize the process – I was often informed that this occurred, but I am not entirely sure how recurrent it was. Overall, couples using clinics seemed to be a minority compared to those operating within what I have called *the informal network*. In this chapter I am presenting material collected in the informal network.
10. A recent study carried out by Gruppo Soggettivita Lesbica (2005) sent out over 3,000 questionnaires to lesbian women, of which 700 were returned: a part of the study concentrates on the desire for motherhood and planning of a family. It offers evidence of such practices and presents both qualitative and quantitative data (see pp. 122–47). The study confirms the data presented here collected during fieldwork a few years earlier. It is therefore particularly interesting as it confirms a growing trend.

project of making a family and achieving conception. With the notion of *partners in asexual conception* I indicate additional levels of involvement where arrangements between couples and donors take different forms and shapes. *Partners in asexual conception*, besides being *known donors*, known contributors of biological material, tend not to disappear from the scene as soon as conception is achieved; they may become more than sporadic visitors to the child and sometimes, and in different degrees, shades and levels of involvement and responsibilities, may take on some form of co-parenting (see also Saffron 1994). When the project is initiated by a gay couple, arrangements are rarely, if ever, made with a surrogate mother in the classical sense, as is the case in the United States or as the press has occasionally reported in the UK. Gay couples usually look for a lesbian couple or for a single lesbian woman willing to share the procreative project. They often explore possibilities for partners in asexual conception and co-parenting with close friends. If not possible or undesirable, they look for potential partners using a go-between with the aim of finding a couple and/or a lesbian woman happy to embark in the procreative project and in some form of co-parenting. For gay men, it is more difficult to initiate such journeys. On the other side, lesbian women and couples depend strongly on gay men to achieve conception.

Orianna (in a relationship with Alessia):

The main difficulty is to establish what kind of family you want to make. If you go for an unknown donor that means that you have already made a clear choice ... you may be able to forget that he ever existed, then you will be able to work out the whole thing as a couple and as mothers.

Manuela (in a relationship with Lorena):

There are all sorts of complications that initially – as soon as you think about having a baby – do not come to your mind. But then suddenly, you realize the huge difference that it is going to make if you choose someone you know or someone you do not.

Manuela's partner, Lorena:

In the end we thought that it's better to keep it open, and accept that we must find ways to include this guy in some way because he is going to be part of this thing. The question is then what kind of role, how involved is he going to be? It's totally artificial to keep him completely out. I would never be able to have a baby without this person – and I would probably never be able to be involved with someone who just dissolves in that way. It's terrible to think that maybe you meet him at the bakery and he is the biological father of the child and you just pretend there is no connection at all.

Gay men are differently placed: their role as donors or possibly partners in asexual conception gives them a more ambivalent status, although not necessarily a powerless one. On the one hand, their position as *initiating couples* is much more tenuous as although they can offer their semen, and can aid conception, they do not easily, if ever, find women and couples willing to relinquish babies after conception. Lesbian women making families by donation wish their children to live with them on a permanent basis. On the other hand, though, they are not left out completely: increasingly lesbian women and couples seem to stress the importance of establishing different kinds of *after-donation, after-conception, after-birth* relations with gay men and couples. Lesbian women are thus increasingly willing to nurture ties of some kind. This gives gay men and couples ample space to negotiate degrees of involvement and exercise some control, besides offering them a chance to initiate the process too.

Tommaso (in a relationship with Alessandro):

For us it was a consequence of thinking what we really wanted together, *as a gay couple*. We wanted to be two fathers giving at the same time a mother to our child. This was always our main concern, especially for me I think. We wanted to find a woman-mother-to-be ready to share fifty-fifty the care and the raising of the baby [although they both would have wished the child to live with them].

Paolo (in a relationship with Stefano):

For a very long time we thought we could only be traditional donors. I donated a few times because a friend of mine asked if I were willing to do it and I did it. This was more than ten years ago. I regret it now that I did it, without asking for some contact with the children. I was young at the time and those were the years that it was really inconceivable. You could not even say you wanted to have a child as a gay man. But to lose any contact in that way, it now seems also inconceivable ... No. I am not going to go back and look for those children, because that was the agreement. But I regret it. I deeply regret it. If they come and look for me, I would be very happy to meet them – to establish something. It would be wonderful if that happened. ... This time is going to be very different with Carlotta and Tatiana. I am a kind of donor but also I will father the baby with my partner in some way. We have not worked out the details yet. It's quite complicated. But I intend to support the baby financially too.

Massimiliano (in a relationship with Alberto):

I was quite explicit that we wanted to be involved with the baby in some way. She initially said no – that she was only looking for a donation – we said that we could not do that way. Six months later she called back, and she said that she wanted to meet us and discuss it. We were thrilled. It was fantastic – it was the most natural conversation we ever had. It was just absolutely OK, apart from when we discussed the last name and she said she would not compromise on her last name. I felt really hurt by that initially. But then I understood it. *This is her own child, I am the biological father. This is not our child. It's different.* So I am OK with it now. (emphasis added)

Lesbian and gay couples making plans, out of multiple choices, show that the project of conceiving a child, contemplating unknown or known donors or partners in asexual conception, has all sorts of potential implications. Conception activates relations that up to that point in time were non-existent; third parties all become implicated in ways that are not temporary and contingent but can be long term and permanent if so wished. I would claim so even in the case of unknown donors, for reasons that will emerge later. By planning and achieving conception outside the 'realm of love and fusion' (Schneider's coinage, 1980 [1968]), lesbian and gay couples do not deny the possibility of making significant ties (see also Weston 1997; Hayden 1995). They are 'rather' imagining possibilities that are complex, intricate and at the same time flexible.

The Lesbian and Gay Way: Practices of Inclusion

Same-sex couples seem to be able to reflect more openly than heterosexual couples attending programmes of gamete donation on the significance and implications of relying on third parties to achieve conception (and it is not surprising, of course, given the symmetrical composition of the couple). They are able to make fully explicit certain social, cultural and, not least, emotional implications that third parties' contribution generates.

Noemi (in a relationship with Sofia):

At first we just wanted an unknown donor, we spent two years just looking for a donor, but each time the go-between told us 'there is this possibility, this man, he is gay, or this man, a friend and so on'. *We just could not cope with not knowing who he was.* It was awkward for both of us – *to make plans about having this intimate link with somebody and not to know his face.* I think especially for us, as lesbians, is really unnatural this male invasion – in the body and as persons. (emphasis added)

Carlotta (in a relationship with Cecilia):

I think that as a lesbian if you suddenly need a man you want to know who he is. At the end of the day you are going to have a baby with him – his stuff is going to make you pregnant. *When you think about it, in this way, I can assure you, that you start to be extremely selective.* For work I meet lots of men. So many times in the past two years just the idea that one of those men could be the one making Cecilia pregnant and me a mother was just too appalling. (emphasis added)

Lesbian women and couples express vividly how important it can be to know who is the other. As Saffron says, many have a 'desire to know what the donor looks like ... women may want to know the donor so to have an image of him as a real person, rather than being totally unknown' (1994: 9). Suddenly, the decision to embark on such projects generates a cascade of effects, including a desire to familiarize with the individual who is making it possible, and without whom a baby could not be conceived. Although third parties, with respect to the initiating couple, do not have emotional significance, they do acquire a significance with respect to the conception project itself. Knowing does matter, as third parties are not perceived as mere genetic material but as whole persons with their own history, background, personality and, not least, 'face'. This is equally emphasized by gay men: familiarity and knowledge of the other is paramount. Gay men and couples wish to know and possibly exercise a choice themselves as well. Federico, in recounting his first meeting to plan conception, acknowledges the power of knowing the future mother of his baby:

> I wanted to see her – the baby would *physically come out from me and her.* When we met I liked her so much, it seemed almost the perfect woman to make a baby with and partly share the growing up. (emphasis added)

Ettore (in a relationship with Giovanni):
> Initially it's all about wanting to know her. It's weird not knowing who she is. *Then when you choose each other,* when you like each other, then you start asking all sorts of other questions: *what kind of relationship are you going to have with a lesbian mother, and as a donor?* (emphasis added)

Lesbian and gay couples envisage and construe ties out of such projects. These take multiple and variegated forms, which often change in time and space, and are highly dependent on circumstances: on the relationships that the initiating couple already has in place, on the role that the non-biological parent will take on (see Hayden 1992), on the demands that partners in asexual conception and donors put on initiating couples, on the presence of other relatives in the life of couples and not least friends who are experiencing similar situations (for different accounts see Lewin 1993; Benkov 1994; Saffron 1994; Ali 1996; Weston 1997). Couples establish different *practices of inclusion,* as I like to call them, whilst experimenting with new forms of living (Weeks et. al. 2001). As the quotes below show, both lesbian and gay couples quickly realize that such projects generate all sorts of pressures as soon as they develop and shape into reality; they give life to a web of uncontemplated relations.

Orianna (in a relationship with Alessia):
> If you do not want to cut across relations *then you have to contemplate everyone in the picture. Then there are at least you, your partner* [the non biological mother], *him* [the biological father] *and the baby. But of course it is never like that really. There are always many more people that you have to think about.* (emphasis added)

Tommaso (in a relationship with Alessandro):
> Then we started to think that the mother-to-be could easily be into another relationship, then we thought about her partner, and the chance that she would be wanting to be a mother as well, and then we started to think about all the potential relatives, selecting on our side who *we really wished to be part of our project, and who we did not, and after we got stuck on the relatives of the mother, and also those of the other mother.* (emphasis added)

Once lesbian and gay couples put in place practices of inclusion, the question is then, what kind of relations should be formed? What should those relations consist of? What makes those relations significant, unique and inalienable? How permanent should they be? How fixed in time and space?

Simona (in a relationship with Chiara):

We have tried to work out as much as possible in detail; but we have also left it quite open. We know that the arrangement can change if things do not work out as we have planned them now. We have kept firm that I will never forbid him to see the child – I believe you can trust each other on that side of things. We all have lots to lose if we mess around with each other.

Simona's partner, Chiara:

Equally, things could go wrong, very wrong, and it would be very messy. We are aware of that – then we would revert to the fact that he is just a *traditional donor* I suppose – It's all quite uncertain I think. (emphasis added)

Paolo (in a relationship with Stefano):

At some point we were thinking of mixing the semen, so that *we could both be fathers at the same time, in the same position towards the child and towards the biological mother.* But then we thought that we should keep it clean for the baby – and the girls also said no way. We would know it anyway who is the baby's father. Mixing the semen also would have created much imbalance with Tatiana [the non-biological mother]. Although if everything goes to plan, she is going to have a baby next – does it sound all crazy?

Paolo's partner, Stefano:

It's a bit crazy if you look at it from a heterosexual point of view – it's all in excess. We are two couples – two families; then we are a couple of biological parents, a couple of non-biological parents, so to speak, that makes four parents with four different parenting roles, but legally only the two of them have parental rights; the two non–biological parents – that is, me and Tatiana, are in a difficult position if something happens – and then there are the grandparents and Paolo's sister for instance – that sort of family people. It's a big family. It's crazy but quite an interesting arrangement, I think. Then if it all goes well and we go ahead with my biological child it will all kind of reverse.

Several lesbian and gay couples also envisage the possibility of operating practices of inclusion postponed in time and space. When they dislike donors' active involvement in their present life, or when donors dislike the idea of being involved, they put in place different kinds of agreements. In these cases lesbian couples secure the potential presence of third parties for the future and are willing to offer them a *non-secret* place. They set out possibilities to incorporate *the other* at a later stage in the life of their children, if the children so wish. They make sure that such possibilities are, in principle, available to them. Equally, gay men or couples, make themselves in principle traceable and identifiable, but stay distant unless the child wishes to know them and actively look for contact.

> Donatella (in a relationship with Samantha):
>
> We wanted a donor, *unknown but traceable*, because we thought that we do not need to know who he is, but the baby [a baby girl] may want to know sooner or later. She will decide when she grows up. Perhaps even at three or four years old. It's unfair to drive children mad by not telling them who is the father. They have the right to know exactly as we have the right not to. (emphasis added)

> Giacomo (in a relationship with Davide):
>
> I am happy to be a known donor. We are actually known donors as a couple. We have biologically fathered two children – but I think both of us are quite happy not to be involved in a more direct way. Maybe when they grow up, if they look for us – than that's OK. I'll have to see what's going to be my reaction to that.

> Simone (in a relationship with Luca):
>
> I agreed to be a known donor, under the condition that the child would not look for me until she is at least six or seven. It sounds harsh I suppose. But when I made the donation I explained that I would not be able to cope with a small child. I know she is well, and I have got pictures of her; Francesca [biological mother] sends me letters from time to time – if she needs anything I help her as I can afford it. There is no problem there. But I cannot cope with a little girl – at this point in my life I cannot go there at all. I actually think that this arrangement suits Francesca and her partner much better.

Lesbian and gay couples recognize and confer a *status of presence* to third parties and acknowledge their genetic contribution, regardless of the arrangement that is going to be set up and the relationship that is going to be established. As we have seen, their existence is made explicit even in cases in which they are kept unknown but traceable or known but not actively involved in a couple' life and in the upbringing of the child. Contrary to the attitude of heterosexual couples seeking relief in the contribution of anonymous donors in gamete donation programmes, lesbian and gay couples do not recur to what I have called *practices of obliteration*. They do not wish to disinvest third parties' contribution of relevance. By contrast, they openly recognize that such a contribution leads to the making of an individual, who then becomes a person tied to other persons in several ways, including biogenetically to both the biological parents. They are then able to qualify what kind of relevance those biogenetic ties have in their eyes.

> Ugo (in a relationship with Marcello):
>
> It seems to me that those ties [*legami*] are important in themselves – disconnected if you like from anything else. It's quite important to know

that they are there, that you have certain connections. You want to know who is your mother and who is your *biological father* – very basic information. It is a point of departure – I would say. Yes, it's a point of departure. (emphasis added)

Marta (in a relationship with Isabella):

It's really about who you are ... Perhaps they are important to me, and I think it's something that deserves some respect because I have been messing around so much in my life. That's a certainty, if you like. It's my right to know who I am – it's my right to know who are my parents – that there is no confusion about that.

Lucetta (in a relationship with Alessandra):

I do not know who's my father – my mother never wanted to tell me and I will never forgive her for that. I am 39 and I am still wondering what does he look like, do I look like him? Do I have this awful character because of him? Why was I denied the knowledge? It's primarily about knowing where you come from and how you are tied to others – even if it is just information you will never use.

Ettore (in a relationship with Giovanni):

It's all about what you inherit genetically I think. And you want to have some information about that. Just as basic information I think, about yourself.

Interestingly though, whilst believing that such ties (*legami* or *legami biologici*) have their own significance because they tell people how they are biogenetically connected (and by implications how they are *not* connected to others), they also believe that such ties do not by default lead to relationships. They only do if made explicit and activated. Biogenetics does not axiomatically forge relations; as Weston points out: 'gay kinship ideologies ... refuse to naturalise familial ties *by equating biogenetic connection with kinship per se*' (1992: 6, emphasis added). Relationships need to be created out of such ties and, as we have seen, this is entirely a matter of willingness and circumstance. Biogenetics can thus act *as a signifier;* it can have cultural –kinship – significance if, and only if, couples decide to attribute to it that significance. Biogenetics can have a transformative power and be a carrier of *forms of permanence*, if couples invest it with agency. This has emerged already in various ways in the previous quotes. In the quotes below it is even more evident.

Lucetta (in a relationship with Alessandra):

When she was born – as soon as she was born – the donor came to visit us, but then he never showed up again. We thought that it was perfect in many ways. We were very keen to be a lesbian family in every sense – but then, at some point, I felt that she [baby girl] also deserved to know her biological father. She started to ask for him and they met. This was not actually the original agreement – as he was not meant to be around this much. But we did not feel like stopping that or interfering. It seemed very cruel.

Ettore (in a relationship with Giovanni):
I do feel that there is something really special between us – I do not see him often at all. I am actually thinking of asking if I can see him more – but it is going to be very difficult. They are both quite strict about the original agreement – but I feel that to know that he is my child has changed completely how I feel about him. If I did not know him, if I did not have any relationship with him, probably I would be fine – it's very difficult – extremely hard – to have to contain the relationship now. I am his father, so it's really hard.

Marta (in a relationship with Isabella):
For four years the donor kept quite distant and we did not do anything to get in touch. We were slightly disappointed because we thought he was a caring guy – he disappeared although he said he was there if we needed him. We left it like that. Then he reappeared and asked to see her [the child]. We explained that he was her biological father – in the back of our mind we always wanted them to meet. She was instinctively thrilled and explained that she had been thinking about her dad 'for many years' but did not ask as she did not want to upset us as she knew we were two mums and could not transform ourselves into a dad. It was heartbreaking.

In making decisions on which relations to forge, and to what extent, lesbian and gay couples find themselves reflecting on the significance that they attribute to other significant ties –entirely non-biological – such as those that link together the non-biological parent to the child as well as all other *kin by inclusion*. They similarly construe and turn on relations which become very significant, long term and invested of permanence – Weston points out, 'in the absence of any axiomatic association of a particular type of social tie... anyone can "work" to turn a relationship into kinship' (1992: 24). Lesbian and gay couples move beyond biogenetics and the relationships that can be forged through biogenetics, and turn social ties into powerful ties. In many cases couples make a great effort to forge social ties with members of the family and relatives where children will grow up.

Ugo (in a relationship with Marcello):
At the same time, of course, it is all relative – because of course your family is almost entirely made up of people who are not biologically related to you. The family is about feelings [*sentimenti*]. The child is loved by both of us, and only one of us is the biological father. The same goes for the relatives.

Lucetta (in a relationship with Alessandra):
She called both of us *mamma* from the very beginning and we did not correct it – when she asked who was the real one we said that we were both real, just that one of us took care of her in the tummy, the other one outside

the tummy – so that she received double care all the way along. She was very satisfied with that, I think, and she continued to call us both *mamma*. We tried to explain that there can be different kinds of mothers and that both can love their children in that special and unique way.

The Lesbian and Gay Way: Practices of Relatedness

The very fact that in the world outside there is no explicit lesbian and gay reproductive and family model to adhere to pushes lesbian and gay couples to rethink and reconceptualize reproductive choices, conception stories, the family project, biological and social relatedness anew. Lesbian and gay couples, contrary to heterosexual couples, cannot afford to – and do not want to – act *normatively*. The decision to make a family and have children does not come as expected. Sharing a procreative and a family project with a same-sex partner is a complicated adventure, which as Weston points out for the American case, often seems to vary from couple to couple (1992: 7). The *initiating couple* orchestrates the project with others outside the couple, makes arrangements, and thinks through the relationships that will be activated after conception and the birth of the baby. A new family configuration needs to be put in place. In particular, the re-elaboration and re-evaluation of the role conferred to unknown/known donors and asexual partners in conception exposes the kinship strategy to which lesbian and gay couples subscribe. It makes explicit lesbian and gay practices of kinship or, better, relatedness, borrowing from Carsten's notion. As Carsten points out, relatedness offers a possibility 'to move away from a *pre-given opposition* between the biological and the social' (2000: 4, emphasis added). Lesbian and gay couples place unknown/known donors and partners in asexual conception in some sort of relationship with the offspring, if not active then potentially so if the child so wishes in the future, the assumption being that, as couples claim, 'the child may at some point in life feel the need to know where it comes from' or 'search for contact' or 'wish to meet the donor'. The contributor of genetic material is, since the early days of planning the pregnancy, thought of and made into a *presence* – even when they remain a stranger to the intentional couples (for instance when transactions rest in the hands of a go-between). Such *practices of inclusion* reflect the relevance assigned to biogenetic ties and confirm the more general point made by Weston that 'gay kinship ideologies have not contested the dominant paradigm of biogenetic kinship by deconstructing biology' (1992: 17). Lesbian and gay couples do not totally depart from a kinship model that incorporates biology; they still rely on it, although in very diverse ways – indeed partly changing its form. In their view, biogenetic ties are not important in absolute terms; it's their knowledge that may be relevant. They have a significance that should not be dismissed a priori. If biogenetic ties are not everything, they are not nothing either. The family that lesbian and gay couples construe, the relations that they envisage – in very different ways from heterosexual couples using programmes of gamete donation – are possible precisely because they allow themselves to incorporate one and the other – biogenetics and the social – without perceiving them as mutually exclusive in the context of donation. The biogenetic tie is understood and conceptualized in the light of the social relations that they wish to be created, or that they believe should be left to be created for their children. Lesbian and gay couples thus generate two synchronic stories whereby biogenetic and social ties are distributed and mobilized between different social actors, in the present or potentially in the future.

CHAPTER 7

THE TRAFFIC IN KINSHIP: SOUTHERN EUROPE AND EURO-AMERICA

Now while I take 'English' as my exemplar of a folk model and thus illustrative of Euro-American kinship thinking, there is also good reason to suppose that *the trivialisation of kinship in social life is a characteristic that may well distinguish it from some continental or Southern European models* (though it may give it affinity to aspects of 'American' kinship). It is of interest insofar as it has helped shape British anthropological theorising on kinship. Both belong to a cultural area I have called 'modernist' or 'pluralist' (Strathern 1992: 106, emphasis added)

Introduction

With *Conceiving Kinship* I have taken a journey into heterosexual couples' discovery of infertility and their subsequent decision to undergo programmes of gamete donation; I have added (in footnotes) the views of couples who also suffer from infertility but undergo programmes in public hospitals with their own gametes, or else adopt. In the same fashion, I have included the views of couples with no vested interest. In the last chapter I have offered an alternative perspective: that of lesbian and gay couples planning families by donation. In all cases gamete donation has remained at the core of the investigation. What I have done is to pirouette around it. This chapter highlights certain key (kinship) notions emerging from the ethnography and sets them side by side with English and American key notions as discussed by Marilyn Strathern (1992a, 1992b, 2005) and David Schneider (1980 [1968]) in their respective work.[1] It is

---

1. It should be added, of course, that Strathern's work on English kinship speaks to Schneider's, as she herself puts it: 'David Schneider is the anthropological father of this book [*After Nature*] since it is both with and against his ideas of kinship that it is written' (1992a: xviii).

not the aim of this chapter to discuss the work of the two scholars in any detail, nor to report on the critique that has been made of Schneider (including his own) for offering too general an account of American kinship (see Yanagisako 1978; Schneider and Smith 1973; Schneider 1980 [1968]). Nor is the aim to discuss more recent developments in kinship studies (see Franklin and McKinnon 2003; Carsten 2000, 2004). The aim is rather to single out those notions that highlight (at one level of analysis) *a continuum* between Italian, English and American kinship.[2] The chapter then continues with a section on the challenges that programmes of gamete donation pose and a section on what I have called the exercise of generalization. It concludes with a note on *Conceiving Kinship* as a whole.

Ethnographic Reflections: Some Key Notions

Conceiving Kinship has brought to light, consistently and recurrently, certain key notions of procreation, the family, the making of kin, biogenetic and social relatedness from multiple perspectives. What has emerged can be summarized as follows. Infertile heterosexual couples attending programmes of gamete donation subscribe heavily to a notion of reproduction described by Delaney as: 'typical in the West, as primarily a natural and therefore universal process' (1986: 495). The reproductive event is seen as the outcome of, or the response to, a *biological drive* which takes over at some point in life, firstly announcing itself as a desire and thereafter as a need. As a biological drive, the reproductive event can be postponed, but not suppressed. As much as it is appropriate to control one's own life, it is *not* appropriate to leave unfulfilled the requirements of mother nature. The notion of reproduction envisions the body as manufactured, predisposed, for the reproductive enterprise. The reproductive body works autonomously – it naturally produces eggs and semen. Any malfunction is fundamentally unnatural – as such infertility is unnatural and needs to be fought against. Similarly infertile heterosexual couples attending programmes with their own gametes or opting for adoption consider the reproductive event, and the reproductive body, *natural* and *naturalize* it but conceptualize both as *inimitable* and *inviolable*. These couples feel that there is a limit to what one can do regardless of how much medical technology is available. The limit is defined by a vision of nature as an intricate milieu of possibilities, and as fundamentally indomitable, not by science and technology as such. Nature not only might impede the reproductive event, and the reproductive body to function efficiently, but it might hinder any attempt to succeed with the aid of technology. Thus, nature works on two levels: that of creating the conditions for successful reproduction and that of

---

2. There are generally few attempts in the anthropological literature to carry out similar exercises – in those cases they tend to highlight certain *discontinuities*: see Bouquet (1993); Kahn (2000).

creating the conditions for obstructing it. Lesbian and gay couples making families by donation also consider reproduction a natural event – a natural occurrence: 'a miracle', 'a beautiful fact of life', as many of them say – but they do not take it for granted (this holds particularly for lesbian women). The reproductive event is thus seen as *natural*, but it is not *naturalized*. It belongs to the domain of nature, but nature is not conceptualized as super-imposing and dictating the rules. Nature is a realm of its own that one can take or leave depending on circumstances.

A second, powerful, idiom is that of the family. Family is, as Dolgin puts it for the American 'rooted in the very nature of things' (1997: 1). Infertile heterosexual couples choosing programmes of gamete donation perceive the family as natural as much as ideal. Indeed, in their view *natural equals ideal*. In the natural/ideal configuration the family consists of a mother and a father, possibly a daughter and a son. As Schneider puts it: '"Family" can mean all of one's relatives, but "my family" or "the family" means a unit which contains a husband and wife and their children, all of whom are kinds of relatives' (1980 [1968]: 30). Or, even more straightforwardly: 'the reproducing pair, living together with their offspring *is* the family' (ibid.: 109). The idiom of the natural/ideal family prevails because it embodies multiple, derivative, sub-idioms: endurance, permanence, solidarity, trust, love, affection, emotional attachment, care and intimacy (Schneider 1980 [1968]: 52). *Fare la famiglia* (to start a family) is a project for life, where parties enter forever relationships despite future life events and circumstances; as Strathern points out: 'the family dissolves but the kinship remains' (2005: 26). Similar powerful idioms are also generated by infertile heterosexual couples attending programmes with their own gametes and by couples choosing adoption - albeit less normatively. For these couples, the family with own children represents a natural/ideal configuration, but it is not the only possible configuration and it is not compelling. The form is not compelling; not only does a couple already make a family – a family is a family with or without children – but a family can be fully formed by adoption. In vivid contrast with couples choosing programmes of gamete donation, couples attending programmes with their own gametes and couples choosing adoption place more emphasis on the partnership, and on what they can think of, imagine, dream, and create together – after the discovery of infertility. They emphasise present possibilities, the family that, at this point in time, can be created despite infertility, rather than what they might have expected. They stress that the sub-idioms of endurance, permanence, solidarity, trust, love, affection, emotional attachment, care and intimacy can be created through other family forms. This is equally so, if not more, for lesbian and gay couples making families by donation. Lesbian and gay couples produce flexible notions of the family and subscribe to multiple representations – it can be said richer representations. As Weston (1997) points out, the family changes and varies from case to case. There is no natural/ideal family configuration: the lesbian and gay family is anything but normative, and it is fully theorized as a choice.

A third powerful idiom is that of biological inheritance. Couples attending programmes of gamete donation, couples attending programmes with their own gametes or else adopting, as well as lesbian and gay couples, invariably acknowledge the role played by both (biological) parents in the making of a child. They all see biogenetic material shared 50 per cent between the biological mother and father. They all visualise such contribution as described by Schneider in *American Kinship*: 'both mother and father give substantially the same kinds and amounts of material to the child and ... the child's whole biogenetic identity or any part of it comes half from the mother, half from the father' (1980 [1968]: 23).

Similarly, all couples seem to conceptualise the biogenetic identity of the child as carrying the biogenetic history of the two (biological) family lines; the significance of biogenetics is not limited to uniting the biological parents to the children, the children to the parents, and eventually the children to each other – but to uniting the children to each family's ancestors. Strathern clarifies the point in relation to the English:

> In popular belief, the parts that an individual person 'gets' from either mother or father may be thought of as parts of other ancestors that 'show' in descending generations. What is visualized, then, is a transmission of substance proportionate to the individual recipient. The individual contains within her or him so much percentage of blood from this or that grandparent, an image I would call literalist. (1992a: 80)

In this capacity, biogenetically related children act as a powerful *trait d'union* across multiple generations. They become a loci of connections: they make visible resemblance and similitude conceptualized in specific physical and, importantly, social traits. This is also observed by Schneider for Americans (1980 [1968]). He says:

> It is a belief in common biological constitution that aspects like *temperament*, build, physiognomy and *habits* are noted as signs of this shared *biological make-up*, this special identity of relatives with each other. Children are said to look like their parents. (ibid.: 25, emphasis added)

Where all couples differ (i.e. couples choosing gamete donation, those using their gametes and lesbian and gay couples) is in the emphasis that they place on each of those constructs. Infertile heterosexual couples undergoing programmes of gamete donation, for instance, conceptualise biogenetic ties as paramount and wish to maximize and preserve what they can. Despite suffering from impaired infertility, they long for an ideal configuration whereby the harmonious correspondence between biogenetic and social ties creates exclusivity. They maximise exclusivity with notions such as 'what one person gets', 'temperament', 'build', 'physiognomy' and 'habits'. Infertile heterosexual couples who undergo programmes with their own gametes or decide to adopt perceive biogenetic ties as equally paramount in forging distinctive social relations but can envisage alternative scenarios (in the former case if the programme does not work) and can contemplate an absent link between biogenetics and social relations. They see biogenic ties as occupying a special place because they *can* generate special connections between (specific) individuals. However, contrary to couples choosing gamete donation, when this is not achievable due to infertility, they seem able to foresee social ties entirely outside the realm of biogenetics. As Weston (1997) has already noticed, lesbian and gay couples go a step further: they fully re-conceptualize biogenetics, without totally departing from it. They see biogenetic ties has having degrees of relevance depending on who they tie and the *potential* social relations that they forge. Such ties are indeed paramount between the birth mother and the child; but also (often) of various significance between donors and the child. With respect to the latter, couples emphasise that biogenetics does not axiomatically generate relations, but can in principle if activated. They claim the value of knowing, transparency and information. Social ties are conferred high relevance with respect to co-mothers and co-fathers, and the respective families' relatives, all of whom need to grow into their roles.

Programmes of Gamete Donation: Challenging (in Principle) the 'Model'

In a programme of gamete donation the entire model of procreation and the family whose parents and children are biogenetically and socially linked, is upset (in principle). It is upset for Italians as well as for Euro-Americans. The notion of a unique pair making a unique individual in the act of conception dissolves, as does the notion of shared substance. It does because, following the model, it is the pair (the intentional couple, the parents to be) that matters. The pair that matters is, of course, the one in which 'the biological' and 'the social' are in 'harmony and in a relation of exclusivity'. In a programme of gamete donation that very unique pair splits precisely in the act of reproducing itself in the child – in a moment of dramatic symbolic importance for Italians as well as for Euro-Americans. The basic axiom that a child comes from a couple made of two individuals in a relationship of uniqueness and intimacy (which presupposes knowledge of each other) is disrupted in all respects. In a programme of gamete donation the child is conceived from two individuals who are socially disconnected and displaced – totally unknown to one another. The procreative project is indeed subdivided between the couple and the anonymous donor and it is dislocated in both time and space. Significantly, it is the absolute lack of relation between parties (couples and donors) and the temporary relations between couples and clinicians that permits all those involved in a programme of gamete donation to operate with opposite intent. If recipients rely on a programme of gamete donation to have their *own* child, donors contribute to the programme for reasons other than to have their child. Clinicians act as providers in the act of conception. By definition, the recipients, donors and clinicians are all very differently placed in the overall enterprise. (Donors in particular are made into very particular type of individuals: they are required to produce a narrative of willing for others to have children, but are regarded with high suspicion if they themselves express the desire to parent.) Moreover, the various strategies that clinics put in place fully show how severely the model is upset (in principle) and how remedies are then sought. The matching donor strategy, for instance, for which the expectation (never a reality, of course) is that the clinic will provide a donor whose genetic material matches the genetic material of the infertile parent, demonstrates the extent to which programmes of gamete donation jeopardize the cultural order. Otherwise, why would a matching donor practice be so highly regarded and advocated?[3] Equally, practices of concealment and practices of obliteration (despite the increasing international trend towards transparency at a policy level),[4] which serve to eliminate third parties from the scenario, make evident the unresolved cultural complexities attached to such programmes.

However, as much as programmes of gamete donation fully upset the model (in principle), their success rests in their ability to recreate the conditions for *replicating the model* (in practice). I have described in this book how this occurs. The programme, which by definition cannot recreate *for real* a relation of exclusivity between the biological and the social parent (or the biogenetically linked family whose children incorporate the combination of both parents' physical similarities and social traits), can instead *simulate* it. The programme can reinstate *a pretence* of shared substance by silently introducing third parties' genetic material; it can fully mask such intrusion, thereby making achievable what would be impossible by other means due to impaired infertility. The programme simulates fertility, reproduction and biological relatedness, creating the perfect conditions for social relatedness. It allows the creation of (what these couples consider to be) the normal family, favouring a scenario where everything happens as it ought to. The couple will have their own child, become parents and start their own family. The family that most nearly resembles the natural/ideal one will be mimicked. This is

---

3. In the United States in particular, in the context of a free and liberalized market in assisted conception services, there is an increasing emphasis on choosing and matching your own donor. At the time of fieldwork a rhetoric of matching donors was reported by couples during interviews (although they could not choose from a catalogue). Clinicians too always pointed out the importance of such practice. As already mentioned, due to restrictive legislative measures passed in 2004, programmes of gamete donation are currently suspended in Italy although there is much pressure for a change in the law from all parties.
4. As mentioned in Chapter 5, legislation towards transparency is increasingly in place in many European countries (including the UK). However, current research shows that couples still tend not to reveal to their children their means of conception if born through programmes of gamete donation.

why a programme of gamete donation, in the view of these couples who have a highly normative view of the model, is the best possible option.

Interestingly, this is also why programmes of gamete donation are less appealing for couples who, in contrast, decide to undergo programmes with their own gametes only and/or opt for adoption. These are couples who equally draw upon the model, but have a less normative view of procreation, the family and biological and social relatedness. These couples are ultimately able to produce more flexible representations and can therefore envisage alterative ways *to make* kinship. They can, for instance, envisage social relatedness in the absence of biogenetic ties between mothers, fathers and children. They can accept being biogenetically unrelated to their children; they can accept not preserving 50 per cent of potential biogenetic ties that are left between the fertile parent and the offspring. They believe that the model is valid only if accomplished in toto, and not just in part. They are able to overcome the desire to reproduce themselves in their own children, can create diverse family forms and achieve relatedness entirely on social ties. Lesbian and gay couples illustrate these points even more blatantly – and in more complex ways. They *remake* kinship by rethinking both biogenetic and social relations. In making families by donation they generate diverse forms of relatedness between biological parents, co-parents, unknown/known donors, partner in asexual conception and other relatives. As Williams says, 'as actors, humans make meanings in culture, stretch meanings to fit circumstances where they no longer work, and build new meanings as well' (1991: 3).

Italian versus Euro-American Kinship: Generalizing the 'Model'

Throughout this work, in bringing to light certain ways of conceptualizing kinship and kin relations, I have worked at more than one level of generalization. On a first level, I have made generalizations about the notions deployed by (Italian) infertile heterosexual couples attending programmes of gamete donation, and I have added the views of infertile heterosexual couples undertaking treatment with their own gametes or else choosing adoption. Following a particular trajectory, I have also made generalizations with respect to couples leading a different lifestyle and claiming diverse worldviews: I have presented the notions lesbian and gay couples put to work in making families by donation. On a second level, throughout the book I have made further generalizations to highlight *continuities* between the Italian data and the literature on assisted conception as well as the literature on lesbian and gay couples making families by donation, typically labelled as Euro-American. At this second level of generalization I have also summarized (in the previous sections) continuities between certain Italian kinship notions as they emerge from the narratives of couples and the work of

Strathern (1992a, 1992b) and Schneider (1980 [1968]), respectively on English and American kinship. Overall I have generalized, but do not intend *to homogenize*. Nor do I intend to suggest that what Italians may share with Euro-Americans in the arena of kinship is necessarily illustrative of other social domains.[5] I wish to make clear that I have confined my generalizations to the domain of kinship (abstractly invoked) but also that I have limited myself to certain notions that permeate conceptions of procreation, the family and relatedness. More to the point: the focus on certain continuities does not imply that Italians and Euro-Americans – whether in the English version (Wolfram 1987; Edwards 2000; Edwards and Strathern 2000; Strathern 1992a, 1992b) or in the North American (Schneider 1980 [1968]; Dolgin 1997) – are alike in the way in which they ultimately make kinship. Let me expand this further. There are reasons to believe that whilst they (conceptually) articulate certain kinship notions drawing on the same *kinship repertoire*, they do not necessarily do more than that. Italians as Euro-Americans *make* kinship in everyday life, in a multitude of ways. We know that this occurs all the time within the same cultural contexts, within the same locales (see D'Andrade and Strauss 1992). We have seen it powerfully in the Italian case: *Italians themselves draw on the same kinship repertoire but make kinship differently in practice*. As Strathern points out, it is well known to social scientists that cultural forms always come with their own peculiarities (Strathern 1992b: 23). It is thus intriguing that the way in which certain kinship notions are (conceptually) articulated does not necessarily tell us the place they occupy and the shape they take in the contingency of people's lives. In other words, to claim that there are powerful continuities in how Italians, like Euro-Americans, make kinship known to themselves is one task, one level of argument, and possibly the first step. A different task is to explore how Italians and Euro-Americans make kinship in every day life drawing upon the same model, in a variety of different milieux. The notions are the same as we have seen, but as they may have multiple manifestations, there is need for more comparative and cross-cultural work at an ethnographic level, as Carsten rightly suggests (2004). We can thus (conceptually) generalize, but cannot homogenize. We can suggest that the Euro-American model goes beyond Euro-America, but the form and shape that it takes, how it manifests itself, cannot be taken too much for granted. In other words, to investigate kinship practices may be a challenging enterprise just as it is to correct a long-standing distance between anthropological cultural areas such as the Euro-American and the Southern European.

---

5. With this remark I do not imply that kinship is a discrete domain (see Collier and Yanagisako 1987: 6). I simply wish to point out here that I do not have data to expand my claim.

A Concluding Note: *Conceiving Kinship*

With *Conceiving Kinship* my aim has been to show that couples relying on the contribution of anonymous donors to achieve conception voice certain narratives that recall a specific kinship model. I have elicited the ways in which a kinship model works as point of reference for couples who conceptualize kinship similarly but *make* it differently, such as couples using their own gametes and couples choosing adoption. And I have – perhaps unusually – suggested a comparison with lesbian and gay couples, who enjoy reproductive capacity but are in a relationship that impedes procreation within, as Schneider puts it, the 'realm of love and fusion' (1980 [1968]). I have detailed how diverse couples, with diverse emotional and sexual orientation, employ certain kinship notions in a place such as Italy, at a particular point in history and in response to some of the possibilities offered by advances in the field of science and medicine. I have argued for a core of idioms with respect to procreation, the family and biological and social relatedness in the context of procreative impediment and partnership differences. What has emerged confirms Strathern's point that 'in effect [people] borrow from what already exists as knowledge. What already exists serves as a "model"; that is a source of description or representation' (1993: 188). Couples conceptualize their own kinship narrative by drawing upon and referring to a known/shared kinship *repertoire*. Within that repertoire, couples create and live their own lives in unique ways. The specificity of social and cultural forms is made up of a multiplicity of representations and, consequently, experiences. For I have suggested that informants who see themselves as inhabiting and belonging to the same social and cultural world (many times they would colloquially say, 'here in Italy!' or 'us Italians!') conceptualize kinship in ways that are complex, articulated, multiple, yet that always depart from some fundamental core idioms. They operate on a model (following Strathern) or a 'cultural system; that is, as a system of symbols' (following Schneider, 1980 [1968]: 1) that always works as a point of reference. The comparison between infertile heterosexual couples choosing a programme of gamete donation or refusing it and adopting instead, and lesbian and gay couples has shown how procreation, family and biogenetic and social relatedness are articulated differently in relation to the particular way couples see, interpret and locate themselves in the world of others. The relevance of biogenetic ties, for instance, is claimed by all of them in quite dissimilar ways and with different levels of emphasis, although in all cases it remains pivotal. If infertile heterosexual couples choosing gamete donation claim the relevance of the genetic tie and maximize it between one parent and the child, claiming that they prefer to settle with what they can preserve (that is 50 per cent of the biological material between the fertile parent and the child – although this comes at the cost of incorporating 50 per cent of anonymous biogenetic material), infertile heterosexual couples rejecting gamete donation claim the relevance of biogenetic ties in the name of an opposite argument:

such ties cannot be re-adjusted to overcome the problem of infertility. Here, biogenetic ties are critical in as much as they tie the child to the wife and husband in *exclusive* ways (and not *selectively* – to one or the other partner only) and in as much as they tie the pair to each other via the offspring. In this view, the biogenetic tie is desirable because it creates exclusive relationships (between parents, children and each family side). In another way altogether, the relevance of the biogenetic tie is claimed in lesbian and gay partnerships, not because it is presumed to tie both parents in a couple to the children (which is an obvious impossibility in same-sex partnerships), and certainly not both parents in a couple to each other, but because it is perceived as *contributing to the personal history of the offspring*. In all cases the biogenetic tie is ultimately regarded for its capacity to generate social identity and social relations.

And I have gone a bit further. Throughout this work I have revealed certain continuities referring to the existing literature on assisted conception and on lesbian and gay couples –entirely informed by a Euro-American perspective – as well as to the specific works on English and American kinship by Strathern (1992a, 1992b) and Schneider (1980 [1968]). Within two anthropologically differentiated and dichotomized fields such as the Euro-American and the Southern European (to be restricted to the Italian case),[6] *continuity and not discontinuity*, is what has emerged. I have addressed the possibility that Euro-Americans and Italians – here as couples with a unique experience of infertility and with diverse emotional and sexual orientation – share, deploy and exploit very similar notions to assemble their own kinship narrative. (They do not necessarily share more than those very specific notions.) In short, I have thus suggested that the Italian data can be read in the light of a conceptual framework called Euro-American rather than Mediterranean. I have suggested the possibility that the Euro-American model contains more than it claims, and that *Conceiving Kinship* has offered some evidence in that direction.

---

6. I would like to limit such a claim to the Italian case and to the ethnographic material I have presented in this work. I have not carried out other work in the Mediterranean area and I have not investigated other domains apart from that of kinship in the specificity of procreation, family, biological and social relatedness.

APPENDIX I

ASSISTED CONCEPTION IN ITALY: A LEGISLATIVE AND POLITICAL CONTROVERSY, 1996 – 99

The account that follows puts the ethnography of assisted conception into its wider legislative and political context, focusing on significant parliamentary events that occurred between 1996 and 1999. As well as describing the political climate at the time of fieldwork, it offers a snapshot of an acrimonious legislative and political fight over the approval of legislation between left and right-wing political parties that is salient in the history of Italian assisted conception. The reader should bear in mind that this is but one instance in a long history of failed attempts to legislate by the Italian parliament, dating back to 1958. This history is one of bitter and contrasting views, in which the position of Italy's main parties has shifted frequently according to transient political swings and moments of opportunism. Yet the failure to pass effective legislation has been constant. The political play on assisted conception conforms to the general pattern described by Spotts and Wieser: 'even though Italian Politics are in a constant state of flux – and almost invariably confound prediction – their basic traits are remarkably constant' (1986: ix). As a result of such repeated failure, assisted conception has long been unregulated in Italy. It was only in 2004, more than a decade later than in many other European countries, that the first law was passed (see Box 8). Yet again this was done in a situation of extreme controversy and tension (see Valentini 2004; Franco 2005) but this time with the result of altering the status quo. The material cited consists largely of legal documents and media cuttings, but on several occasions, opinions and interpretations of events have been backed up by informants working on national commissions and for national newspapers.

Towards a Unified Text:
Political Controversies over Legislation

The long history of attempts to legislate in the field of assisted conception in Italy has been documented elsewhere (see Di Pietro and Casini 2002).[1] Up to 1996, throughout several Administrations (*Legislature*), over twenty bills (*Proposte di Legge*) were presented by various interested parties but these were either rejected soon after submission or never reached approval in both Chambers, as required to complete the parliamentary *iter* and become law.[2] Amongst this history of failures a significant series of events occurred between 1996 and 1999 (which partly overlaps with the time of fieldwork) around the *unified text* presented by the Commission of Social Affairs of the Chamber of Deputies *(Commissione Affari Sociali della Camera dei Deputati)*. In 1996 the Commission was given the governmental mandate to revise seven bills on assisted conception (previously presented to the Chamber) and work on a *unified text* to be called 'Dispositions in the Field of Assisted Conception' (*Disposizioni in Materia di Procreazione Assistita*). Deputy Marida Bolognesi, member of the *Democratici di Sinistra* party (new reformed left-wing party), was appointed to chair the Commission; the other members mirrored the plurality of political representation in government and opposition. The Commission worked on the *unified text* for over a year. On 29 October 1997, when the text was ready for submission to the Chamber of Deputies, many members asked for last-minute amendments; the final version could only be approved on 27 January 1998. After various heated discussions two articles were added: article 21, which exempted clinicians who 'conscientiously objected' to providing treatment; and article 5, which banned the use of donated gametes by same-sex couples and singles. In short, the *unified text* legitimated assisted conception for couples of opposite sex, married or in a permanent relationship with an age limit of 52. It allowed treatment with donated gametes (both egg and sperm donation) as long as the donation of gametes was voluntary, unremunerated and anonymous and as long as no more than five live births were obtained from the same donor. The *unified text* also banned any experimentation on

---

1. Di Pietro and Casini (2002), for instance, report that the first bill was presented in 1958 and proposed imprisonment for women who had undergone artificial insemination by donor, as well as for their husbands who agreed to such treatment and the donors who provided the semen. They report that in the subsequent administration in 1963 a similar text was proposed but, like the first, it never reached parliamentary discussion (2002: 617).
2. In Italy a bill (*a proposta or disegno di legge*) can be presented by different bodies: the government, regions, CNEL and citizens. Permanent Commissions nominated by the parliament (the Chamber of Deputies – *Camera dei Deputati* – or the Senate – *Il Senato della Repubblica*) have the mandate to evaluate proposals before discussion and approval. As Italy has a bicameral system any submitted bill needs to pass and be fully approved by the two Chambers in order to become law and be enforced.

embryos (unless requested for diagnostic-therapeutic ends), cloning, prenatal screening and selection of embryos with eugenic intent, ectogenesis and any surrogacy arrangement. So formulated, the *unified text* – called *Disegno di Legge Bolognesi* – was presented nationally, via the media, as the best possible political compromise that Catholic and non-Catholic, left and right-wing members of the Commission were able to reach. In response many in the Chamber of Deputies – mostly exponents of the right-wing and centre parties – expressed their disappointment with the outcome of the work of the Commission and claimed the unconstitutionality *(inconstituzionalità)* of the *unified text*. This allegation instantly led to the involvement of the Commission of Constitutional Affairs and Justice of the Chamber of Deputies *(Commissione Affari Costituzionali e Giustizia della Camera)*, which has the specific mandate of judging the constitutionality of bills (that is, whether, a bill does or does not infringe any article of the Italian Constitution, *La Costituzione della Repubblica Italiana* 1947). There were two contentious points: the legality (legitimacy) of treatment requiring the use of third parties gametes in view of article 2 which protects personal identity. It was claimed that *'l'eterologa'* (artificial insemination by donor) *does not protect the individual from the right to be born with a defined biogenetic identity;* and the legality of granting unmarried couples access to a treatment that would favour the birth of children outside marriage. This latter point, judged by many as even more provocative and factious than the previous, was made in view of article 29, which encourages the birth of children within the family 'as the natural society based on marriage'.[3] Professor of Law Ferrando commented at the time:

> It is true that under art. 29 of the Italian Constitution the Family is defined as being based on marriage, but it is also true that art. 2 *safeguards any social group where human personality is developed*. Indeed the same art. 2 clearly states that *the family is not safeguarded as an institution, whose interests are greater than those of an individual*. Therefore those who are safeguarded are the members of the social group; thereby upholding the rights of individuals inside the family. (unpublished paper, emphasis added)

With 19 votes in favour of bill's constitutionality and 13 against it[4] the Commission of Constitutional Affairs and Justice of the Chamber of Deputies, headed by Deputy Rosa Russo Jervolino (interestingly, a supporter of the Catholic centre party *Partito Popolare Italiano* and an openly

---

3. Some Catholic politicians, such as Carlo Giovanardi of *CCD (Centro Cristiano Democratici)*, went so far in asserting the unconstitutionality of the bill that he declared: '[the bill] is not only in contrast with the articles of the Constitution but more fundamentally with *the principle* of the Constitution' (*Avvenire*, 3 June 1998: 3, emphasis added).
4. The 19 votes in favour of the bill's constitutionality were expressed primarily by those belonging to the major coalition forming the government, thus the centre-left-wing party l'Ulivo. The 13 votes against were expressed by the opposition, il Polo formed by *Alleanza Nazionale (reformed ex-fascist party)*,

practising Catholic herself), concluded that the *unified text Disegno di Legge Bolognesi* was constitutional (2 June 1998).[5] This made the bill formally ready to be presented, discussed, amended if necessary by the Chamber of Deputies and, if approved, ready to go to the Senate, for the second round of discussions, amendments and possible final approval.

At this point, the forces that disapproved of the *unified text* and that failed in their attempt to declare it unconstitutional, including members of the Chamber of Deputies and political party spokespeople, put in place a new strategy. They severely attacked the President of the Constitutional Commission Rosa Russo Jervolino for failing to sink the text there and then. She was attacked widely and nationally, via the media, by the Catholic mainstream and by her own party. Amongst various accusations she was blamed for voting against her own Catholic principles. A supporter of a right-wing political party (*Alleanza Nazionale*) commented:

This is a vote which contradicts the Catholic principle; it contradicts the respect of the human being because it allows artificial procreation between unmarried couples. It also contradicts the Constitution in relation to the values of marriage ... and family. (*La Repubblica*, 3 June 1998: 10)

President Rosa Russo Jervolino was also severely opposed by the President of the Constitutional Court (*Corte Costituzionale*), which is the highest institutional body able to assess the constitutionality of a bill and which is independent of parliament. The president, Antonio Baldassarre, strongly believed in the unconstitutionality of the *unified text* because, from his point of view, it introduced a new concept of the family altogether. He stated: 'here there is an attempt to introduce with an ordinary *law an altogether new juridical concept of the family*' (emphasis added). He also authoritatively added: 'it [*Commissione Affari Costituzionali*] expressed a political vote, the *Consulta* [*Corte Costituzionale*] will be always able to assert the contrary'. (*Avvenire*, 3 June 1998: 2)

Interestingly, the question of unconstitutionality was raized for diametrically opposite reasons by the more liberal lay side too. Stefano Rodotà, a well-known constitutionalist, Professor of Law at the University La Sapienza in Rome and President of the National Commission for the Protection of Privacy (*Commissione per la Tutela della Privacy*), invoked the unconstitutionality of the *unified text* because of the denial of medical treatment to couples and individuals on the basis of their civil status: being in a couple or being single. He said:

This is a text which shows the attempt to take over the control of women's reproductive capacity: this is a prohibitionist bill ... if assisted conception is considered a therapy against infertility, to discriminate against single women is unconstitutional on the base of art. 3 [of the Constitution] because it denies the right to health [*il diritto alla salute*] on the basis of a personal condition such as to be single. (*L'Unità*, 4 June, 1998)

Questions of unconstitutionality were so contested that the *unified text* could not be presented to the Chamber of Deputies as initially planned. On

9 June 1998 the President of the Commission Marida Bolognesi, given the extreme political tensions surrounding the presentation of the text, announced 'a pause for reflection'. In other words, she postponed the parliamentary debate to a more promising political time, in the hope that there would be one. Instead, this marked the beginning of an even more controversial period for law making on assisted conception. Amidst all sorts of political accusations and sudden changes of position a new flow of arguments took over.

After several months of conflicts, disagreements, public debates and more formal 'pauses of reflections' – in a political moment which turned out to be no more favourable – on 3 February 1999, the *unified text* reached the Chamber of Deputies. In two days the bill was completely sunk – unexpectedly at this point, as many hoped that a political compromise would be finally reached. A cross-party majority (*una maggioranza ad hoc con voto trasversale*)[6] voted against the *unified text*. It was a vote with a boomerang effect: it sank the bill and any hope of legislation on assisted conception. This created strong dissent within several political parties (mostly within right-wing and centre parties such as *Forza Italia* and *Popolari*), and caused (two days later, 5 February 1999) President Marida Bolognesi's resignation. A week later (12 February 1999) a successor president was formally appointed by Bolognesi. Bizarrely, she had to appoint Deputy Alessandro Cè of the right-wing party *Lega Nord* who led the major controversy and created an ad hoc coalition to vote against her bill. There was very little that Bolognesi could have done to save almost three years of hard work and careful political negotiations. It was immediately clear, as at other times in the past and as in the Italian political tradition, that the reasons for failure were far greater than the *unified text* on assisted conception itself – the text was an excellent political target, a political commodity (it was highly contentious and legislation was long awaited in the country) for a much wider political project. As the intellectual Italian establishment widely commented at the time, the opposition aimed for the dismissal of the current centre-left government in power headed by Massimo D'Alema. To sink Bolognesi's *unified text* was a glamorous act in that direction.

The Death of the Unified Text: The Rise of a New Controversy

When the text *Disegno di Legge Bolognesi* – which from most quarters was believed to be the best possible in the circumstances – was definitively killed off, the Commission, now headed by the right-wing Deputy Alessandro Cè started to work on amendments as requested by the cross-party coalition (*coalizione trasversale*). The coalition introduced, amongst others, the

---

6. The cross-party majority was created by a sudden coalition of right-wing supporters and Catholics – mostly by party members forming the opposition (*Il Polo*) and by party members forming the coalition of the government (I Popolari).

following substantial prohibitions: a ban on any form of treatment with donated gametes to unmarried couples; a ban on the cryopreservation of embryos; and a limit on embryo implantation to three at a time.[7] Together with the crux of the original *unified text* rewritten, an unexpected proposal was also put forward: it was suggested that all existing frozen embryos (which had been estimated to be approximately 80,000 to 90,000) be declared 'adoptable' – biological parents were required to decide within two years either for implantation or for adoptability. The text also imposed severe penalties, including imprisonment, for those who did not follow the new rules.

The revised text became formally ready for discussion – and in principle approval – by the Chamber of Deputies at the end of February 1999, but at this point a major national controversy arose. The battle between the governing coalition and the opposition, including the Roman Catholic Church, became crude[8] and the entire political scene reversed once again. A new group of *dissidents*, representing different political right-wing positions, surprisingly associated themselves with the major left-wing force of the country (still forming the government) and, suddenly formed a new coalition. Amongst the new dissidents were female politicians who, in the name of a battle against what was considered the most restrictive law of assisted conception in Europe, created the *cross-party women's alliance*, as it came to be defined by the media.[9] This was a political move to momentarily oppose the bill – but of course it created new fractures in the main national right and centre-wing parties (e.g. *Forza Italia, Lega Nord* and even the Catholic *Partito Popolare Italiano*) and, more generally, severe fractures for the broad right – centre coalition (at the time, the opposition).

At this point politicians claimed it was impossible to reconcile private and public, personal and political positions. They made a formal and open distinction between what they called 'individual conscience' (*coscienza*

---

7. The ban on cryopreservation has severe consequences for women who have to repeat the whole treatment in case of unsuccessful implantation as they have no surplus embryos from previous treatments that can be implanted. Deputy Buffo of left-wing party Democratici di Sinistra said at the time: 'This is a decision which puts women's health at risk ... it risks to make of assisted conception a via cruces for women, children and clinicians.' On the same occasion Deputy Bolognesi added: 'In this way embryos will be more protected than women, with the risk of leaving private clinics the chance to speculate even more on the length of treatments and complexity of therapies' (*Corriere della Sera*, 24 March 1999: 19).
8. Unfortunately it is not possible to reproduce in the text the intensity of the controversy that raged from this point in time onwards due to the number of politicians involved and the never-ending contradictory statements that were made.
9. Amongst the women politicians who briefly associated themselves with the political agenda of the left-wing coalition there was, for instance, Deputy Mussolini, the niece of Duce Mussolini, supporter of the ex-fascist party *Alleanza Nazionale*, definitely a major case of right-wing political orientation.

individuale) and 'political stands' (*appartenenza politica*) and unanimously announced that they would vote for a bill on the basis of the former and not the latter (*La Repubblica*, 25 February 1999: 2). Assisted conception, they declared, profoundly touched and meddled with personal values, beliefs and worldviews. No one could be asked to act without taking into consideration the personal in such a morally overloaded arena. With such a premise the newly formed cross-party women's alliance, to whom other deputies associated themselves, initiated a new discussion in the Chamber of Deputies (25 February 1999).[10] The completely revised bill under Deputy Alessandro Cè (*ex-unified text Disegno di Legge Bolognesi*) was discussed all over again, article by article, and new changes were made: access to unmarried couples but in stable and permanent relationship was granted, and the age limit was withdrawn, making it possible to reformulate article 5 which was particularly contentious. The rewritten article now stated: 'Couples of age [*maggiorenni*], of opposite sex, married or cohabiting, of an age potentially fertile [can access treatment to assisted conception]' (*La Repubblica*, 26 February 1999: 10). However, it was impossible to reach any further agreement on the ban on treatment with third parties' gametes, nor on cryopreservation or any form of embryo research. Despite the initiative and the careful negotiation of the *newly formed cross-party women's alliance* the bill remained extremely restrictive – an example of a heavy bill (*legge pesante*) rather than of a light one (*legge leggera*). As such, there was little hope that it would be approved in the second Chamber, the Senate, which had a stronger centre-left formation.[11] At that point, everyone knew that once again Italy had failed, after over three years of negotiations (and over more than twenty years of previous attempts) to legislate in the field of assisted conception. It was clear to everyone that it would be for the next government to take up the challenge again.

The Political Project behind Assisted Conception: 1996–99

The controversies around the approval of legislation represent a point in history, but, as stated at the outset, are representative of what happened

---

10. As I have already pointed out, I am unable to detail the long and intricate history of legislative attempts and the long-standing pattern of repetition. However, I may be able to offer here an example: on this occasion (exactly as before) the issue of unconstitutionality came up once again. As soon as the discussion started, Deputy Cananzi of centre-party *Partito Popolare Italiano* stated: 'Assisted conception outside marriage? Unconstitutional!'. This again led to a new controversy which filled up the pages of the media for weeks and weeks all over again.
11. As mentioned, in order to become national law (*Legge dello Stato*) the bill has to be discussed and approved by the Senate – *Il Senato della Repubblica* – where at that particular point in time the left-wing coalition was much stronger.

several times up to 2004, when for the first time legislation on the matter was approved. The heated debates around assisted conception – and specifically about who should be permitted treatment, the use of third parties' gametes and the number of embryos to be created during each cycle and implanted – reveal and represent the highest point of fracture between secularism (what Italians call *il pensiero laico*) and Catholicism (*il pensiero cattolico*). Both factions, the Catholic and the secular – made up of pluralistic political coalitions – were in a fight they could not afford to lose. In particular the Roman Catholic Church and associated political parties saw assisted conception and the powerful issues that it raised as an opportunity to regain symbolic power which had been damaged and weakened since the main Catholic party, the Christian Democrats (*Democrazia Cristiana*) had lost credibility due to scandals, corruption and long-standing crisis in the late 1980s/beginning of 1990s. Relatively new political parties (*Lega Nord, Forza Italia, Partito Popolare Italiano* – to cite but a few) with their short history in the Italian political scenario had come to be perfect instrumental players. They needed to gain and consolidate power quickly and fitted very well into the Roman Catholic Church's plan to create new coalitions. They were apt to contribute to the manipulation of controversial issues: the legislation over assisted conception was one of those and a perfect one. The issues that assisted conception generated were those that Catholic parties anywhere would want to monopolize. Paradoxically, the most conservative force in the country was precisely the one that contributed most of all to the creation and reproduction of a legislative hole. Catholicism acted as a powerful political force to block every bill that was submitted for discussion and approval and that did not comply with the Catholic radical *credo* and preaching.[12] The Catholic media – for example the daily voice of national newspapers such as *L'Osservatore Romano* and *L'Avvenire* – and centre and right-wing political parties all came together to play a tough game. In such a context, the left-wing main party *Democratici di Sinistra*, which was governing the country at the time together with minor left-wing parties such as *Rifondazione Comunista* and national feminist movements, could not but politicize assisted conception, although for diametrically opposite reasons. They had to oppose such a force from the right – of course they spectacularly failed to do so and in 2001 the centre-right-wing coalition headed by Silvio Berlusconi won the election. In a matter of little more than a year the government of Berlusconi passed the most restrictive law on assisted conception in Europe.

---

12. This shows the paradox of Catholicism: such politics resulted in many infertile couples undergoing infertility programmes without the protection of a legislative framework.

BOX 8

The First Italian Law on Assisted Conception, 40/2004

In 2001, the newly elected centre-right government headed by Silvio Berlusconi asked the Commission of Social Affairs of the Chamber of Deputies to examine sixteen bills on assisted conception presented by different political parties. As in the past, the Commission worked on a unified text which was ready in March 2002. As in the past, the bill went through a tortuous process of parliamentary controversies and fights, several rounds of negotiation and the deployment of political forces around the country – but contrary to the past, legislation was approved by the parliament with 277 votes in favour and 222 against. The new legislation, 'Norms in the Field of Assisted Conception' (40/2004) (*Norme in Materia di Procreazione Medicalmente Assistita*), which came into place on 19 February 2004, bans access to treatment for reasons other than diagnosed clinical infertility (i.e. couples who suffer from unexplained infertility are excluded from treatment as well as couples who are carriers of a genetic disorder); it bans any form of treatment with donated gametes, cryopreservation of embryos, embryo experimentation, therapeutic cloning, both in the public and in the private sector. It only allows three embryos to be created at a time, all of which must be implanted regardless of their state of health (even if graded of poor quality).[13] In June 2005 a referendum to modify the most controversial articles was instituted with the intent to reverse the ban on embryo research, cryopreservation, prenatal screening, the obligation to implant all embryos at once, and, importantly, treatment with donated gametes. However, it had to be called void, as the quorum of 50+1 per cent of the electorate was not reached due to low attendance (26 per cent only).[14] Various interpretations have been put on this by commentators: some have suggested that the low attendance was due to the heavy politicking and manoeuvring of the Roman Catholic Church (Pope Benedict XVI endorsed the call from bishops to boycott the referendum)[15] and to similar pressure from the centre-right government of Berlusconi, which was responsible in the first place for passing the law. Many commentators also

---

13. This clause contravenes law 194/1978 on abortion – in theory the successfully implanted low-graded embryo can be thereafter legally removed under such law.
14. It has been reported that 74.1 per cent of Italians did not vote – but it is estimated that 30 per cent would not have voted anyway (so called 'physiological abstention'). A great majority of those who voted (26 per cent) chose to vote 'yes' (77 – 88 per cent) that is, to modify the bans and limits of the law (see Doe et al. 2005 for a brief summary).
15. Although the Roman Catholic Church played a strong role in inviting Italians to abstain, it must be said that other reasons may have counted in the decision too. In earlier referenda the intervention of the Roman Catholic Church failed miserably: in 1974 the referendum on divorce saw 87.7 per cent of the overall electorate expressing a vote with 59.6 per cent voting to maintain the law (Law 898/1970); in 1981 the referendum on abortion saw 79.4 per cent of the overall electorate expressing a vote with 88.4 per cent voting to maintain the law (Law 194/1978).

BOX 8: *continued*

observed the influential role of the media controlled (overwhelmingly) by Berlusconi[16] or attributed failure to a widespread political apathy. Whatever the reasons for the poor turn out – the referendum was void and none of the articles of the 40/2004 law were reversed or modified.

To date, the law remains one of the most restrictive in Europe. It is ironic (and unfortunate for infertile couples) that the law is a complete reversal of the previous situation: it has turned Italy from the 'Wild West of Assisted Conception' – from the land of inward 'reproductive tourism' – into one where couples must go abroad to seek assistance. It regulates the provision of treatment and services, it makes clinics and clinicians accountable and it has stopped an unregulated market in reproduction and infertility, but it is so rigid that it forbids the provision of the most basic services and treatment; services that are readily available outside Italy. It is argued that it dangerously denies basic civil liberties to women (Franco 2005: 39) and it contradicts pre-existing legislation on abortion (law 194/1978) and the Italian Constitution (article 3, 1947). As Rodatà points out, it is a law that has created a legislative 'involution' (2004).

---

16. This view finds support in the work of Comstock and Sharrer (2005), who claim: 'research indicates that people discount their own opinions and experiences in favour of those of experts as espoused in the media. The framing of news coverage thus has a profound impact on public opinion …'. However, many scholars argue that the relationship between the media and reception processes is far more complex (see Hall 1980, 1997; Morely 1992, 1995; Askew 2002).

APPENDIX II

Profile of Infertile Heterosexual Couples

This appendix contains a brief biographic descriptions of couples interviewed.

Anna and Paolo

Anna, 34 years old, from Naples, university degree in literature, teacher. Paolo, 40 years old, from Naples, university degree in political sciences, academic. Catholic sometimes practising. Married since 1992.

Marta and Massimo

Marta, 29 years old, from Aosta, university degree in architecture, architect. Massimo, 35 years old, from Milan, university degree in public relations, manager in a publishing company. Catholic but non-practising. Married since 1994.

Irene and Alessandro

Irene, 27 years old, from Palermo, college qualification (*magistrali*), primary school teacher. Alessandro, 31 years old, from Bergamo, vocational qualification, salesman. Catholic but non-practising. Married since 1993.

Sofia and Filippo

Sofia, 32 years old, from Milan, college qualification (*liceo*), secretary. Filippo, 37 years old, from Milan, college qualification (*liceo*), computer officer. Catholic sometimes practising. Married since 1992.

Barbara and Giancarlo

Barbara, 38 years old, from Rimini, college qualification (*liceo*), shop owner. Giancarlo, 40 years old, from Rimini, college qualification (*liceo*), shop owner. Catholic but non-believers. Married since 1990.

Gioia and Leopoldo

Gioia, 32 years old, from Messina, college qualification (*liceo*), primary school teacher. Leopoldo, 31 years old, from Chiavari, short degree in economics, fashion shop owner. Catholics but non-practising. Married since 1994.

Nadina and Ottavio

Nadina, 32 years old, from Piacenza, college qualification (*liceo*), housewife, Ottavio 32 years old, Pavia, college qualification (*liceo*), family firm owner. Catholics but non-believers. Married since 1990.

Costanza and Antonio

Costanza, 34 years old, from Varese, college qualification (*liceo*), housewife since treatment began. Antonio, 35 years old, from Rome, college qualification (*liceo*), family firm owner. Catholics but non-believers. Married since 1988.

Maddalena and Pietro

Maddalena, 32 years old, from Turin, university degree in history, housewife. Pietro, 30 years old, from Milan, university degree in law, lawyer. Catholics sometimes practising. Married since 1993.

Mariangela and Luca

Mariangela, 33 years old, from Verona, degree in fashion, fashion designer. Luca, 36 years old, from Rome, university degree in law, lawyer. Catholics sometimes practising. Married since 1993.

Marina and Michele

Marina, 40 years old, from Bari, college qualification (*liceo*), nursery teacher. Michele, 41 years old, from Rome, college qualification (*liceo*), computer graphic artist. Catholics but non-believers. Married since 1988.

Michela and Daniele

Michela, 26 years old, from Florence, college qualification (*liceo*), concierge. Daniele, 29 years old, from Florence, college qualification (*liceo*), taxi driver. Catholic sometimes practising. Married since 1996.

Veronica and Giacomo

Veronica, 39 years old, from Venice, college qualification (*liceo*), administrator. Giacomo, 40 years old, from Pordenone, college qualification (*liceo*), family firm owner. Catholics sometimes practising. Married since 1992.

Lucia and Daniele

Lucia, 37 years old, from Trento, college qualification (*liceo*), secretary. Davide, 37 years old, from Trento, university degree in economics, bank manager. Catholics and practising. Married since 1991.

Olga and Teodoro

Olga, 37 years old, from Florence, university degree in maths, teacher. Teodoro, 38 years old, from Pisa, university degree in biology, manager in a pharmaceutical company. Catholics but non-believers. Married since 1993.

Matilde and Vittorio

Matilde, 29 years old, from Padova, college qualification (*liceo*), librarian. Vittorio, 36 years old, from Bologna, college qualification (*liceo*), salesman. Catholics but non-believers. Married since 1994.

Penelope and Andrea

Penelope, 38 years old, from Milano, university degree in psychology, child psychologist. Andrea, 38 years old, from Turin, short degree in art, manager in a art gallery. Catholics sometimes practising. Married since 1990.

Elisabetta and Germano

Elisabetta, 35 years old, from Milan, college qualification (*liceo*), secretary. Germano, 35 years old, from Bergamo, college qualification (*liceo*), accountant. Catholics but non-practising. Married since 1992.

Caterina and Edoardo

Caterina, 29 years old, from Milano, college qualification (*liceo*), travel agent. Edoardo, 30 years old, from Bergamo, college qualification (*liceo*), travel agent. Catholics but non-practising. Married since 1994.

Federica and Gianluca

Federica, 32 years old, from Pavia, college qualification (*liceo*), saleswoman. Gianluca, 38 years old, from Pavia, vocational qualification, computer officer. Catholics sometimes practising. Married since 1989.

Cinzia and Michelangelo

Cinzia, 35 years old, from Novara, college qualification (*liceo*), secretary. Michelangelo, 35 years old, from Novara, college qualification (*liceo*), bank manager. Catholics but non-practising. Married since 1993.

Matilde and Giorgio

Matilde, 28 years old, from Udine, college qualification (*liceo*), model. Giorgio, 34 years old, from Udine, university degree in languages, translator. Catholics but non-practising. Married since 1996.

Letizia and Roberto

Letizia, 32 years old, from Turin, college qualification (*liceo*), secretary. Roberto, 33 years old, from Turin, vocational qualification, factory worker. Catholics but non-practising. Married since 1995.

Emma and Guido

Emma, 34 years old, from Viterbo, college qualification (*liceo*), secretary. Guido, 39 years old, from Florence, college qualification (*liceo*), small family firm owner. Catholics sometimes practising. Married since 1989.

Stefania and Mattia

Stefania, 31 years old, from Cremona, vocational qualification, sales assistant. Mattia, 40 years old, from Cremona, college qualification (*liceo*), small family firm owner. Catholics sometimes practising. Married since 1992.

Sonia and Eugenio

Sonia, 36 years old, from Rome, short degree in public relations, public relations manager. Eugenio, 34 years old, from Genova, college qualification (*liceo*), accountant. Catholics but non-believers. Married since 1990.

Francesca and Marco

Francesca, 31 years old, from Bologna, short degree in nursing, nurse. Marco, 33 years old, from Bologna, college qualification (*liceo*), bank clerk. Catholics but non-believers. Married since 1990.

Margherita and Pier

Margherita, 36 years old, from Bologna, college qualification (*liceo*), family firm owner. Pier, 35 years old, from Bologna, college qualification (*liceo*), photographer. Catholics but non-believers. Married since 1994.

Giulia and Fabrizio

Giulia, 35 years old, from Venice, college qualification (*liceo*), sport teacher. Fabrizio, 42 years old, from Naples, college qualification (*liceo*), chairman in a multinational company. Catholics but non-practising. Married since 1987.

APPENDIX IIA

PROFILES OF LESBIAN AND GAY COUPLES

Nora and Cristiana

Nora, 37 years old, from Naples, university degree in medicine, dentist, has lived with Cristiana for four years. Cristiana, 37 years old, from Naples, college qualification (*liceo*), assistant dentist. Catholics but non-belivers.

Lorena and Manuela

Lorena, 34 years old, from Milan, college qualification (*liceo*), saleswoman, has lived with Manuela for two years. Manuela, 36 years old, from Pavia, university degree in geology, geologist in a multinational petrol company. Catholics but non-practising.

Maura and Silvia

Maura, 34 years old, from Como, college qualification (*liceo*), a policewoman, has lived with Silvia for six years. Silvia, 29 years old, from Milan, short degree in physiotherapy, physiotherapist. Catholics but non-practising.

Micaela and Lucia

Micaela, 45 years old, from Turin, university degree in economics, manager in a marketing company, has been with Lucia for four years. Lucia, 38 years old, from Turin, university degree in literature, teacher. Catholics but non-practising.

Sebastiano and Giulio

Sebastiano, 35 years old, from Rome, university degree in maths, secondary school maths teacher, previously married, divorced after a year of marriage, has now been living with Giulio for seven years. Giulio, 33 years old, from Palermo, college qualification (*liceo*), fashion salesman. Catholics but non-believers.

Andrea and Enrico

Andrea, 35 years old, from Cagliari, university degree in medicine, dentist. He has been living with Enrico for three years. Enrico, 36 years old, from Cagliari, university degree in economics, financial advisor. Catholics but non-believers.

Gigi and Carlo

Gigi, 36 years old, from Perugia, college qualification (*liceo*), consultant. He has been living with Carlo for three years. Carlo, 36 years old, from Taranto, college qualification (*liceo*), consultant. Catholics but non-practising.

Francesca and Susanna

Francesca, 30 years old, from Genova, college qualification (*liceo*), secretary, has lived with Susanna for five years. Susanna, 33 years old, from Genova, college qualification (*liceo*), primary school teacher. Catholics sometimes practising.

Stefania and Caterina

Stefania, 40 years old, from Bari, short degree in speech therapy, speech therapist, has lived with Caterina for ten years. Caterina, 39 years old, from Milan, short degree in nursing, nurse. Catholics sometimes practising.

Paola and Elena

Paola, 29 years old, from Vicenza, university degree in information technology, computer graphic artist, has lived with Elena for seven years. Elena, 30 years old, from Vicenza, short degree in fashion, fabric designer. Catholics but non-believers.

Bruno and Maurizio

Bruno, 29 years old, from Ventimiglia, short degree in arts, designer, has lived with Maurizio for a year. Maurizio, 34 years old, from Venice, college qualification (*liceo*), secretary in a media company. Catholics but non-practising.

Claudio and Marco

Claudio, 30 years old, from Naples, vocational qualification, secretary in a hospital, has lived with Marco for three years. Marco, 32 years old, from Naples, college qualification (*liceo*), factory worker. Catholics but non-believers.

Carolina and Tania

Carolina, 32 years old, from Catania, college qualification (*liceo*), hostess, has lived with Tania for three years, now with a one-year-old child. Tania, 29 years old, from Perugia, college qualification (*liceo*), airport assistant. Catholics but non-believers.

Alessia and Orianna

Alessia, 39 years old, from Milan, university degree in languages, secondary school teacher, has lived with Orianna for five years. Orianna, 39 years old, from Milan, university degree in literature, editor. Catholics but non-believers.

Tommaso and Alessandro

Tommaso, 36 years old, from Catania, university degree in philosophy, head librarian, has been living with Alessandro for two years. Alessandro, 32 years old, university degree in history, manager in a publishing company. Catholics but non-believers.

Paolo and Stefano

Paolo, 36 years old, from Taranto, university degree in medicine, doctor, has been living with Stefano for three years. Stefano, 32 years old, university degree in medicine, manager in a pharmaceutical company. Catholics but non-believers.

Massimiliano and Alberto

Massimiliano, 34 years old, from Vercelli, university degree in biology, researcher, has been living with Alberto for five years. Alberto, 32 years old, from Aosta, university degree in veterinary, vet. Catholics but non-believers.

Noemi and Sofia

Noemi 32 years old, from Padova, college qualification (*liceo*), secretary, has lived with Sofia for seven years. Sofia, 32 years old, from Milan, college qualification (*liceo*), saleswoman. Catholics sometimes practising.

Carlotta and Cecilia

Carlotta, 37 years old, from Milan, short degree in languages, manager in a translating company, has lived with Cecilia for six years. Cecilia, 31 years old, from Milan, short degree in arts, photographer. Catholics but non-practising.

Federico and Andrea

Federico, 36 years old, from Parma, college qualification (*liceo*), sales assistant, has lived with Andrea for one year. Andrea, 35 years old, from Milan, college qualification (*liceo*), administrator. Catholics but non-practising.

Ettore and Giovanni

Ettore, 39 years old, from Milan, college qualification (*liceo*), actor, has lived with Giovanni for four years. Giovanni, 35 years old, from Milan, college qualification (*liceo*), computer graphic designer. Catholics but non-practising.

Simona and Chiara

Simona, 32 years old, from Milan, college qualification (*liceo*), journalist, has lived with Chiara for four years. Chiara, 34 years old, from Milan, college qualification (*liceo*), secretary in a media company. Catholics but non-practising.

Donatella and Samantha

Donatella, 27 years old, from Pordenone, vocational qualification, secretary, has lived with Samantha for four years, they have a child together. Samantha, 26 years old, from Trento, secondary school, bus driver. Catholics but non-believers.

Marta and Isabella

Marta, 30 years old, from Florence, college qualification (*liceo*), secretary, has lived with Isabella for three years. Isabella, 31 years old, from Pisa, college qualification (*liceo*), saleswoman. Catholics but non-believers.

Lucetta and Alessandra

Lucetta, 39 years old, from Rome, university degree in architecture, architect, has lived with Alessandra for seven years, they have a child together. Alessandra, 37 years old, from Rome, university degree in law, lawyer. Catholics but non-believers.

BIBLIOGRAPHY

Abu-Lughod, J. (1991) 'Going Beyond Global Babble'. In A. King, *Culture, Globalisation and the World System*. London: Macmillan.

Ahmed, A. and C. Shore (1995) *The Future of Anthropology: Its Relevance to the Contemporary World*. London: Athlone.

Ali, T. (1996) *We Are Family: Testimonies of Lesbian and Gay Parents*. London and New York: Cassell.

Allum, P. (1990) 'Uniformity Undone: Aspects of Catholic Culture in Post-war Italy'. In Z. Baranski, R.Lumley, *Culture and Conflict in Post-war Italy: Essays on Mass and Popular Culture*. Basingstoke: Macmillan in Association with the Graduate School of European and International Studies, University of Reading.

Anderson, B. (1991) *Imagined Communities: Reflections on the Origin and Spread of Nationalism*. London: Verso.

Appadurai, A. (1996) *Modernity at Large: Cultural Dimensions of Globalization*. Minneapolis: University of Minnesota Press.

Appel, W. (1976) 'The Myth of The Jettatura'. In Maloney, C. *The Evil Eye*. New York: Columbia University Press.

Arditti, R., R. Duelli Klein and S. Minden (1989) *Test-Tube Women: What Future for Motherhood?* London: Pandora.

Arlacchi, P. (1983) *Mafia, Peasants and Great Estates: Society in Traditional Calabria*. Cambridge: Cambridge University Press.

Askew, K. and R. Wilk (2002) *The Anthropology of Media: A Reader*. Oxford Blackwell.

Augè, M. (1995) *Non-Places: Introduction to an Anthropology of Super Modernity*. London: Verso.

Avvenire. (1998) 'Siamo Fuori Dalla Costituzione'. 3 June, p. 2.

Banfield, E. (1958) *The Moral Basis of a Backward Society*. New York: Free Press.

Barbagli, M. and C. Asher (2001) *Omosessuali Moderni. Gay e Lesbiche in Italia*. Bologna: Il Mulino.

Barbagli, M., M. Castiglioni and G. Dalla Zuanna (2003) *Fare Famiglia in Italia. Un Secolo di Cambiamenti*. Bologna: Il Mulino.

Barrett, R. (1978) 'Village Modernisation and Changing Nicknaming Practises in Northern Spain', *Journal of Anthropological Research*, 34: 92–108.

Barret, R.L. and B.E. Robinson (2000) *Gay Fathers: Encouraging the Hearts of Gay Dads and their Families*. San Francisco: Jossey-Bass.

Becker, G. (2000) *The Elusive Embryo: How Women and Men Approach New Reproductive Technologies*. Berkeley: University of California Press.

Bell, R.M. (1979) *Fate and Honor, Family and Village: Demographic and Cultural Change in Rural Italy Since 1800*. Chicago: University of Chicago Press.

Belmonte, T. (1979) *The Broken Fountain*. New York: Columbia University Press.

Benkov, L. (1994) *ReinventingtThe Family: The Emerging Story of Lesbian and Gay Parents*. New York: Crown.

Berg, M. and A. Mol (1998) *Differences in Medicine. Unravelling Practices, Techniques and Bodies*. Durham, NC: Duke University Press.

Bestard-Camps, J. (1991) *What's in a Relative? Household and Family in Formentera*. Oxford: Berg.

Bharadwaj, A. (2000) 'How Some Indian Baby Makers are Made: Media Narratives and Assisted Conception in India'. *Anthropology and Medicine*, 7(1): 63–78.

——— (2002) 'Conception Politics: Medical Egos, Media Spotlights, and the Contest over Test-Tube Firsts in India'. In Inhorn, M. et al. (2002) *Infertility Around The Globe: New Thinking on Childlessness, Gender and Reproductive Technologies*. Berkeley: University of California Press.

——— (2003) 'Why Adoption is not an Option in India: The Visibility of Infertility, the Secrecy of Donor Insemination, and other Cultural Complexities'. *Social Science and Medicine*, 56: 1867–80.

Blok, A. (1975) *The Mafia of a Sicilian Village: A Study in Violent Peasant Entrepreneurs*. Oxford: Basil Blackwell.

Bonaccorso, M. (1994) *Mamme e Papa' Omosessuali. Primo Saggio Italiano sulla Famiglia Omosessuale*. Roma: Editori Riuniti.

——— (2004a) 'Programmes of Gamete Donation: Strategies in (Private) Clinics of Assisted Conception'. In M. Unnithan, *Reproductive Agency, Medicine and the State: Cultural Transformations in Childbearing*. Oxford: Berghahn Books.

——— (2004b) 'Making Connections: Family and Relatedness in Clinics of Assisted Conception in Italy', *Journal of Modern Italy*, 9(1): 59–68.

Bonaccorso, M. (forthcoming) 'The Trivialization of Social Anthropology: An Ethnographic Account'. In S. Coleman and P. Collins *Dislocating Anthropology?: Bases of Longing and Belonging in the Analysis of Contemporary Societies*. Cambridge: Cambridge Scholars Press.

Bono, P. and S. Kemp, (1991) *Italian Feminist Thought: A Reader*. Oxford: Basil Blackwell.

Borghi, L. and A. Taurino (2005) 'Coniugalitá e Generativitá nelle Coppie Omosessuali'. In L. Fruggeri, *Diverse Normalitá*. Roma: Carocci.

Bottino, M. and D. Danna (2004) *La Gaia Famiglia*. Trieste: Asterios Editore.

Bouquet, M. (1993) *Reclaiming English Kinship: Portuguese Refractions on English Kinship Theory*. Manchester: Manchester University Press.

Bourdieu, P. (1992) *Language and Symbolic Power*. Cambridge: Polity Press.

Bozett, F.W. (1987) *Gay and and Lesbian Parents*. New York: Praeger.

Braidotti, R. and N. Lykke (1996) *Between Monsters, Goddesses and Cyborgs: Feminist Confrontations with Science, Medicine and Cyberspace*. London: Zed.

Brandes, S.H. (1975) 'The Structural and Demographic Implications of Nicknames in Navanogal Spain', *American Ethnologist*, 2: 138–48.

——— (1980) *Metaphors of Masculinity: Sex and Status in Andalusian Folklore*. Philadelphia: University of Pennsylvania Press

Brogger J. and D. Gilmore (1997) 'The Matrifocal Family in Iberia: Spain and Portugal Compared', *Ethnology*, 36.

Butler, J. (1999) *Gender Trouble: Feminism and the Subversion of Identity*. New York: Routledge.

Calhoun, C. (2003) *Feminism, the Family and the Politics of the Closet: Lesbian and Gay Displacement*. Oxford University Press.

Cambrosio, A., A. Young, A. and M. Lock (2000) *Living and Working with the New Medical Technologies: Intersections of Inquiry*. Cambridge: Cambridge University Press.

Campbell, J. and J. de Pina-Cabral (1992) *Europe Observed*. Basingstoke: Macmillan Press.

Comstock, G. and E. Sharrer (2005) *The Psychology of Media and Politics*. Elsevier Academic Press.

Cannell, F. (1990) 'Concepts of Parenthood: The Warnock Report, The Gillick Debate, and Modern Myths', *American Ethnologist*, 17(4): 667–86.

Carsten, J. (2000) *Cultures of Relatedness: New Approaches to the Study of Kinship*. Cambridge: Cambridge University Press.

—— (2004) *After Kinship: New Departures in Anthropology*. Cambridge: University Press.

Clifford, J. and G. Marcus (1986) *Writing Culture: The Poetics and Politics of Ethnography*. Berkeley, Los Angeles and London: University of California Press.

Cohen, A. (1994) *Self Consciousness: An Alternative Anthropology of Identity*. London: Routledge.

Cohen, E. (1977) 'Nicknames, Social Boundaries, and Community in an Italian Village', *International Journal Contemporary Sociology*, 14: 101–13.

Cole, J.W. (1977) 'Anthropology Comes Part-Way Home: Community Studies in Europe', *Annual Review of Anthropology*, 6: 349–78.

Cole, J.W. and E. Wolf (1974) *The Hidden Frontier: Ecology and Ethnicity in an Alpine Valley*. New York: Academic Press.

Cole, S. (1991) *Women of the Praia*. Princeton: Princeton University Press.

Collier, J.F. (1997) *From Duty to Desire*. Princeton: Princeton University Press.

Collier, J. and S. Yanagisako (1987) *Gender and Kinship: Essay Towards a Unified Analysis*. Stanford: Stanford University Press.

Connell, R. (1987) *Gender and Power: Society, the Person and Sexual Politics*. Cambridge: Polity Press.

Corea, G. (1985) *Man-Made Women: How New Reproductive Technologies Affect Women*. London: Hutchinson in Association with the Explorations in Feminism Collective.

—— (1987) *The Mother Machine: Reproductive Technologies From Artificial Insemination To Artificial Wombs*. London: Women's Press.

Corea, G. et al. (1987) *Man-Made Women: How Reproductive Technologies Affect Women*. Bloomington and Indianapolis, Indiana University Press.

Cowan, J. (1991) *Dance and the Body Politic in Northern Greece*. Princeton: Princeton University Press.

Cussins, C. (1998a) 'Producing Reproduction: Techniques of Normalization and Naturalisation in Infertility Clinics'. In S. Franklin and H. Ragoné, H. *Reproducing Reproduction: Kinship, Power, and Technological Innovation*. Philadelphia: University of Pennsylvania Press.

—— (1998b) 'Ontological Choreography: Agency for Women Patients in an Infertility Clinic'. In M. Berg and A. Mol, *Differences in Medicine: Unraveling Practices, Techniques and Bodies*. Durham, NC. and London: Duke University Press.

D'Andrade, R. and C. Strauss (1992) *Human Motives and Cultural Models*. Cambridge: Cambridge University Press.

Daniels, K. (1995) 'Information Sharing in DI – A Conflict of Needs and Rights', *Cambridge Quarterly of Healthcare Ethics*, 4: 217–24.
―――― (1998) 'The Semen Provider'. In K. Daniels and E. Haimes, *Donor Insemination: International Social Science Perspectives*. Cambridge: Cambridge University Press.
Daniels, K. and E. Haimes (1998) *Donor Insemination: International Social Science Perspectives*. Cambridge: Cambridge University Press.
Davis, J. (1973) *Land and Family in Pisticci*. London: Athlone Press.
Davis-Floyd, R. and J. Dumit (1998) *Cyborg Babies: From Techno-Sex to Techno-Tots*. New York: Routledge.
Delaney, C. and S. Yanagisako (1995) *Naturalizing Power: Essays in Feminist Cultural Analysis*. London: Routledge.
De Pina-Cabral, J. (1986) *Sons of Adam, Daughters of Eve: The Peasant Worldview of the Alto Minho*. Oxford: Blackwell.
―――― (1989) 'The Mediterranean as a Category of Regional Comparison: A Critical View', *Current Anthropology*, 30(3): 399–406.
Di Pietro, A. and P. Tavella (2006) *Madri Selvagge*. Torino: Einaudi.
Di Pietro, M.L. and M. Casini (2002) 'Procreazione Assistita/Dibattiti Parlamentari', *Medicina e Morale*, 4: 617–66.
Directorate General For Research (1992) *Bioethics in Europe*. European Parliament.
Doe, N., J. Oliva and C. Cianitto (2005) 'Medically Assisted Procreation in Italy: The Referendum and the Roman Catholic Church'. Online: Http://Www.Ccels.Cf.Ac.Uk/Literature/Issue/2005/Doeolivacianitto
Dolgin, J. (1990) 'Status and Contract in Surrogate Motherhood: an Illumination of the Surrogacy Debate', *Buffalo Law Review*, 38: 515–50.
―――― (1997) *Defining the Family: Law, Technology, and Reproduction in an Uneasy Age*. London: New York University Press.
Douglass, W. (1984) *Emigration in a South Italian Town: and Anthropological History*. New Brunswick, NJ: Rutgers University Press.
Du Boulay, J. (1974) *Portrait of a Greek Mountain Village*. Oxford: Clarendon Press.
―――― (1984) 'The Blood: Symbolic Relations Between Descent, Marriage, Incest Prohibitions and Spiritual Kinship in Greece', *Man*, 19: 533–56.
Dubisch, J. (1986) *Gender and Power in Rural Greece*. Princeton: Princeton University Press.
Duelli Klein, R. (1989) *Infertility: Women Speak out about their Experiences of Reproductive Medicine*. London: Pandora.
Edwards, J. (1995) 'Imperatives To Reproduce: Views From Nortwest England on Fertility in The Light of Infertility'. In R. Dunbar, *Reproductive Decisions: Biological and Anthropological Perspectives*. London: Macmillan.
Edwards, J. (1998) 'Donor Insemination and 'Public Opinion'. In K. Daniels and E. Haimes, *Donor Insemination: International Social Science Perspectives*. Cambridge: Cambridge University Press.
―――― et al. (1999 [1993]) *Technologies of Procreation: Kinship in the Age of Assisted Conception*. London: Routledge.
―――― (1999 [1993]) 'Explicit Connections: Ethnographic Enquiry in Northwest England'. In J. Edwards et al., *Technologies of Procreation: Kinship in The Age of Assisted Conception*. London: Routledge.
―――― (2000) *Born and Bred: Idioms of Kinship and New Reproductive Technologies in England*. Oxford: Oxford University Press.

Edwards, J. and M. Strathern (2000) 'Including Our Own'. In J. Carsten, *Cultures of Relatedness: New Approaches to the Study of Kinship*. Cambridge: Cambridge University Press.

Esposito, N. (1989) *Italian Family Structure*. New York: Peter Lang.

Fabian, J. (1983) *Time and the Other*. New York: Columbia University Press.

Ferrando, G. (1998) 'Assisted Reproduction in Italy. The Current Legal Situation', Unpublished paper.

Filippucci, P. (1992) 'Presenting the Past in Bassano: Locality and Localism in a Northern Italian Town', Ph.D. Dissertation, University of Cambridge.

——— (1996) 'Anthropological Perspectives on Culture in Italy'. In D. Forgacs and R. Lumley *Italian Cultural Studies: an Introduction*. Oxford: Oxford University Press.

Finkler, K. (2000) *Experiencing the New Genetics: Family and Kinship on the Medical Frontier*. Philadelphia: University of Pennsylvania Press.

Forgacs, D. and R. Lumley (1996) *Italian Cultural Studies: An Introduction*. Oxford: Oxford University Press.

Franco, V. (2005) *Bioetica e Procreazione Assistita. Le Politiche della Vita tra Libertá e Responsabilitá*. Roma: Donzelli.

Franklin, S. (1997) *Embodied Progress: A Cultural Account of Assisted Conception*. London: Routledge.

Franklin, S. and M. McNeil (1988) 'Review Essay: Recent Literature and Current Feminist Debates on Reproductive Technologies', *Feminist Studies*, 14: 545–60.

Franklin, S. and H. Ragoné (1998) *Reproducing Reproduction: Kinship, Power and Technological Innovation*. Philadelphia: University of Pennsylvania Press.

Franklin, S. and S. Mckinnon (2003) *Relative Values: Reconfiguring Kinship Studies*. Durham, NC: Duke University Press.

Frith, L. (2001) 'Gamete Donation and Anonymity: The Ethical and legal Debate', *Human Reproduction*, 16(5): 818–24.

Freeman, M. (ed) (1996) *Children's Rights: A Comparative Perspective*. Aldershot: Dartmouth Publishers.

Galt, A. (1991) *Far From the Church Bells: Settlement and Society in an Apulian Town*. Cambridge: Cambridge University Press.

Gambetta, D. (1996) *The Sicilian Mafia: The Business of Private Protection*. Cambridge, MA: Harvard University Press.

Garrison, V. et al. (1976) 'The Evil Eye: Envy or Risk of Seizure?' In C. Maloney, *The Evil Eye*. New York: Columbia University Press.

Gay-Y-Blasco, P. (1999) *Gypsies in Madrid*. Oxford: Berg.

Gell, A. (1999) 'Introduction: Notes on Seminar Culture and Some Other Influences'. In E. Hirsch, *The Art of Anthropology: Essays and Diagrams*. London: Athlone.

Gillespie, P. (1999) *Love Makes a Family: Portraits of Lesbian, Gay, Bisexual and Transsexual Parents and their Families*. Amherst: University of Massachusetts Press.

Gilmore, D.D. (1982) 'Anthropology of The Mediterranean Area', *Annual Review of Anthropology*, 11: 175–05.

——— (1987) *Honour and Shame and the Unity of the Mediterranean*. A Special Publication of the American Anthropological Association, N.22. Washington DC: American Anthropological Association.

——— (1990) *Manhood in the Making: Cultural Concepts of Masculinity*. London and New Haven: Yale University Press.

Ginsborg, P. (2003) *Italy and Its Discontents: Family, Civil Society, State: 1980–2001*. London: Penguin Books.

Glover, J. et al. (1989) *Fertility and the Family: The Glover Report on the Reproductive Technologies to the European Commission*. London: Fourth Estate.

Goddard, V. (1996) *Gender, Family and Work in Naples*. Oxford: Berg.

Gottlieb, C. et al. (2000) 'Disclosure of Donor Insemination to the Child: The Impact of Swedish Legislation on Couples' Attitudes', *Human Reproduction*, 15: 2052–56.

Green, S. (1997) *Urban Amazons: Lesbian Feminism and Beyond in the Gender, Sexuality and Identity Battles of London*. Basingstoke: Palgrave Macmillan.

Gringrich, A. and R. Fox (2002) *Anthropology, By Comparison*. London: Routledge.

Gruppo Soggettivitá Lesbica (2005) *Cocktail D'Amore: 700 e Piú Modi di Essere Lesbica*. Roma: Derive Approdi.

Gunning, J. and V. English (1993) Human in Vitro fertilisation: A case Study in regulation of medical Innovation. Aldershot: Dartmouth Publishers.

Gupta, A. and A. Ferguson (1997) *Anthropological Locations: Boundaries and Grounds of a Field Science*. Berkley and Los Angeles: University of California Press.

Haimes, E. (1990) 'Recreating The Family? Policy Considerations Relating to the 'New' Reproductive Technologies'. In M. McNeil et al. *The New Reproductive Technologies*. London: Macmillan.

――― (1992) 'Gamete Donation and the Social Management of Genetic Origins'. In M. Stacey, *Changing Human Reproduction*. London: Sage Publications.

――― (1993) 'Issues of Gender in Gamete Donation', *Social Science and Medicine*, 36(1): 85–89.

Hall, S. (1980) 'Encoding/decoding'. In S. Hall *Culture, Media, Language: Working Papers in Cultural Studies*. London: Hutchinson.

――― (1997)*Representation: Cultural Representations and Signifying Practices*. London: Sage Publications.

Handwerker, L. (2002) 'The Politics of Making Modern Babies in China: Reproductive Technologies and the New Eugenics'. In M. Inhorn et al. (2002) *Infertility Around the Globe: New Thinking on Childlessness, Gender and Reproductive Technologies*. Berkeley: University of California Press.

Hannerz, U. (1996) *Trans-National Connections: Culture, People, Places*. London: Routledge.

Hastrup, K. and K. Olwig (1996) *Siting Culture: The Shifting Anthropological Object*. London: Routledge.

Hauschild, T. (1995) 'E. De Martino e il Postmoderno Antropologico'. Unpublished paper.

Hayden, C. (1992) 'Making Kinship Trouble: Lesbian Motherhood and the Dynamics of Choice in the "America Family"'. Dissertation, University of Virginia.

――― (1995) 'Gender, Genetics, and Generation: Reformulating Biology in Lesbian Kinship', *Cultural Anthropology*, 10(1): 41–63.

Herzfeld, M. (1981) 'Meaning and Morality: A Semiotic Approach to Evil Eye in a Greek Village', *American Ethnologist*, 8: 560–74.

――― (1983) 'Interpreting Kinship Terminology: The Problem of Patriliny in Rural Greece', *Anthropological Quarterly*, 56: 157–66.

――― (1987) *Anthropology Through The Looking-Glass: A Critical Ethnography in the Margins of Europe*. Cambridge: Cambridge University Press.

Hirsh, E. (1999 [1993]) 'Negotiated Limits: Interviews in South-East England'. In Edwards et al., *Technologies of Procreation: Kinship in the Age of Assisted Conception*. London: Routledge.
Hobsbawm, E. and T. Ranger (1983) *The Invention of Tradition*. Cambridge: Cambridge University Press.
Hughes, D. (1978) 'From Brideprice to Dowry in Mediterranean Europe', *Journal of Family History*, 3: 262–96.
Human Fertilisation and Embryology Act (1990). London: HMSO.
Il Corriere Della Sera. (1999) 'Trapianti e Fecondazione Nuovo Stop'. 24 March, p.19.
Inhorn, M. (2003) *Local Babies, Global Science*. London: Routledge.
—— (2005) 'Religion and Reproductive Technologies: IVF and Gamete Donation in the Muslim World', *Anthropology News*, 46(2).
—— (2006) 'He Won't Be My Son': Middle Eastern Muslim Men's Discourses of Adoption and Gamete Donation', *Medical Anthropology Quarterly*, 20 (1): 94–120.
Inhorn, M. and F. Van Balen (2002) *Infertility Around the Globe: New Thinking on Childlessness, Gender and Reproductive Technologies*. Berkeley: University of California Press.
Iszaevich, A. (1980) 'Household Renown: The Traditional Naming System in Catalonia', *Ethnology*, 19: 315–26.
Jackson, A. (1987) *Anthropology at Home*. London: Tavistock Publications.
Just, R. (2000) *A Greek Island Cosmos*. Oxford: James Curry.
Kahn, S. (2000) *Reproducing Jews: A Cultural Account of Assisted Conception in Israel*. Durham, NC: Duke University Press.
Kertzer, D. (1980) *Comrades and Christians: Religion and Political Struggle in Communist Italy*. Cambridge: Cambridge University Press.
—— (1996) *Politics and Symbols: The Italian Communist Party and the Fall of Communism*. London: Yale University Press.
Kertzer, D. and P. Saller (1991) *The Family in Italy: From Antiquity to the Present*. New Haven: Yale University Press.
Klock, S. et al. (1994) 'A Prospective Study of Donor Insemination Recipients: Secrecy, Privacy and Disclosure', *Fertility and Sterility*, 62: 477–84.
Konrad, M. (1998) 'Ova Donation and Symbols of Substance: Some Variations on the Theme of Sex, Gender and the Partible Body', *Man*, (4)4: 643–67.
—— (2005) *Nameless Relations: Anonymity, Melanesia and Reproductive Gift Exchange Between British Ova Donors and Recipients*. Oxford: Berghahn Books.
La Repubblica. (1998) 'I Duellanti della Provetta'. 3 June, p. 10.
—— (1999) 'Camera, Via Libera alle Coppie di Fatto'. 25 February, p. 2.
—— (1999) 'Fecondazione Assistita Cade il Limite di Età'. 26 February, p. 10.
—— (2000) 'La Fecondazione Naufraga al Senato'. 22 June, p. 22.
Lasker, J. (1998) 'The Users of Donor Insemination'. In K. Daniels and E. Haimes, *Donor Insemination: International Social Science Perspectives*. Cambridge: Cambridge University Press.
Lewin, E. (1993) *Lesbian Mothers: Accounts of Gender in American Culture*. Ithaca: Cornell University Press.
Llobera, L. (1986) 'Fieldwork in Southwestern Europe: Anthropological Panacea and Epistemological Straitjacket?', *Critique of Anthropology*, 6(2): 24–33.
Loizos, P. and E. Papataxiarchis (1991) *Contested Identities*. Princeton: Princeton University Press.

Lumley, R. (1990) *States of Emergency: Cultures of Revolt in Italy from 1968 to 1978.* London: Verso.
Lumley, R. and J. Morris (1997) *The New History of the Italian South: The Mezzogiorno Revisited.* Exeter: University of Exeter Press.
L'Unità. (1998) 'É Incostituzionale Negare la Provetta ai Single'. 4 June.
Macdonald, S. (1993) *Inside European Identities: Ethnography in Western Europe.* Oxford: Berg.
Marcus, G. (1995) 'Ethnography in/of The World System. The Emergence of Multi-sited Ethnography', *Annual Review of Anthropology*, 24: 95–117.
―――― (1998) *Ethnography Through Thick and Thin.* Chichester: Princeton University Press.
Marcus, G. and M. Fischer (1986) *Anthropology as Cultural Critique: An Experimental Moment in the Human Sciences.* Chicago: University of Chicago Press.
Martinelli, A., A. Chiesi and S. Stefanizzi (1999) *Recent Social Trends in Italy 1960–95.* Montreal and Kingston: McGill University Press.
McNeil, M., I. Varcoe, and S. Yearley (1990) *The New Reproductive Technologies.* London: Macmillan.
Miller, D. (1995) *Acknowledging Consumption: A Review of New Studies.* London: Routledge.
Miller, R.A. (1974) 'Are Familists Amoral? A Test of Banfield's Amoral Familism Hypothesis in a South Italian Village', *American Ethnologist*, 1: 515–35.
Mitchell, J. (2002) *Ambivalent Europeans: Ritual, Memory and the Public Sphere in Malta.* London: Routledge.
―――― (2002) 'Modernity in the Mediterranean', *Special Issue of Journal of Mediterranean Studies*, 12(1).
Mol, A. (2000) 'Pathology and the Clinic: An Ethnographic Presentation of Two Atherosclerosis'. In Lock, M. et al. *Living and Working with the New Medical Technologies.* Cambridge: Cambridge University Press.
Morely, D. (1992) *Television Audiences and Cultural Studies.* London: Routledge.
―――― . (1995) 'Theories of Consumption in Media Studies'. In D. Miller *Acknowledging Consumption.* London: Routledge.
Moss, L. et al. (1976) 'Mal'occhio, Ayin Ha Ra, Oculus Fascinus, Judenblick'. In C. Maloney, *The Evil Eye.* New York: Columbia University Press.
Nielsen, L. (1996) 'Procreative Tourism, Genetic Testing and the Law'. In N. Lowe and G. Douglas, *Families Across Frontiers.* Netherlands: Kluwer Academic Publishers.
Novaes Bateman S. (1986) 'Semen Banking and Artificial Insemination by Donor in France: Social and Medical Discourse', *Journal of Technology Assessment in Health Care*, 2(2): 219–29.
―――― (1989) 'Giving, Receiving, Repaying: Gamete Donors and Donors Policies in Reproductive Medicines', *Journal of Technology Assessment in Health Care*, 5(4): 639–57.
―――― (1998) 'The Medical Management of Donor Insemination'. In K. Daniels and E. Haimes, *Donor Insemination: International Social Science Perspectives.* Cambridge: Cambridge University Press.
Okely, J. (1996) *Own Or Other Culture.* London: Routledge.
Ong, A. and S. Collier (2005) *Global Assemblages.* Oxford: Blackwell Publishing.
Orobitg, G. and C. Salazar (duke special 2005) 'The Gift of Motherhood: Egg Donation in a Barcelona Infertility Clinic', *Ethnos*, 70 (1): 31–52.
Owen, R. (1997) 'Clinics Shut Over Allegations of Infected Sperm', *Times of London*, 1 December.

Pardo, I. (1996) *Managing Existence in Naples: Morality, Action, and Structure.* Cambridge: Cambridge University Press.

—— (2000) 'When Power Lacks Legitimacy: Relations of Politics and Law to Society in Italy'. In I. Pardo, *Morals of Legitimacy: Between Agency and System.* Oxford: Berghahn Books.

Passerini, L. (1994) 'The Interpretation of Democracy in Italian Women's Movement of the 1970s and 1980s'. *Women's Studies International Forum,* 17(2/3): 23–40.

Paterlini, P. (2004) *Matrimony Gay: Dieci Storie di Famiglie Omosessuali.* Torino: Einaudi.

Peristiany, J. (1965) *Honour and Shame.* London: Weidenfeld and Nicolson.

—— (ed.) (1976) *Kinship and Modernization in Mediterranean Society.* Rome: The Centre for Mediterranean Studies.

Pies, C. (1985) *Considering Parenthood: A Workbook for Lesbians.* San Francisco: Spinters/Aunt Lute.

Pitt Rivers, J. (1965) 'Honour and Social Status'. In J. Pitt Rivers *Honour and Shame.* London: Weidenfeld and Nicolson.

Plesset, S. (2006) *Sheltering Women: Negotiating Gender and Violence in Northern Italy.* Stanford, CA: Stanford University Press.

Polikoff, N. and D. Vaughn (1987) *Politics of the Heart.* Ithaca, NY: Firebrand Books.

Pratt, J. (1996) 'Catholic Culture'. In D. Forgacs and R. Lumley *Italian Cultural Studies: An Introduction.* Oxford: Oxford University Press.

Price, F. (1992) 'Having Triplets, Quads or Quins: Who Bears the Responsibility?'. In M. Stacey *Changing Human Reproduction* London: Sage Publications.

—— (1995) 'Conceiving Relations: Egg and Sperm Donation in Assisted Conception'. In A. Bainham, D. Pearl and R. Pickford *Frontiers of Family Law.* Chichester: Wiley.

—— (1996) 'Now You See It, Now You Don't: Mediating Science and Managing Uncertainty in Reproductive Medicine'. In A. Irwin and B. Wynne *Misunderstanding Science?: The Public Reconstruction of Science and Technology.* Cambridge: Cambridge University Press.

Rabinow, P. (1996) *Essays on the Anthropology of Reason.* Princeton: Princeton University Press.

Ragoné, H. (1994) *Surrogate Motherhood: Conception in the Heart.* Boulder, co: Westview Press.

—— (1996) 'Chasing the Blood Tie: Surrogate Mothers, Adoptive Mothers and Fathers', *American Ethnologist,* 23(2): 352–65.

Rapp, R. (2000) *Testing Women, Testing the Foetus.* New York: Routledge.

Rivière, P. (1985) 'Unscrambling Parenthood: The Warnock Report', *Anthropology Today,* 4: 2–7.

Rogers, S. (1991) *Shaping Modern Times in Rural France.* Princeton: Princeton University Press.

Rossi Barilli, G. (1999) *Il Movimento Gay in Italia.* Milano: Feltrinelli.

Rowland, R. (1992) *Living Laboratories: Women and Reproductive Technologies.* Bloomington: Indiana University Press.

Saffron, L. (1994) *Challenging Conceptions. Planning a Family by Self-Insemination.* London: Cassell.

Saint-Cassia, P. with Constantina Bada (2006) *The Making of the Modern Greek Family.* Cambridge: Cambridge University Press.

Sanders, P. (1995) 'A Missed Appointment: Ernesto De Martino and the Anthropology of the United States'. Unpublished paper.

Saraceno, C. (2003) *Mutamenti della Famiglia e Politiche Sociali in Italia*. Bologna: Il Mulino.
Saraceno, C. and M. Naldini (2001) *Sociologia della Famiglia*. Bologna: Il Mulino.
Saunders, G. (1984) 'Contemporary Italian Cultural Anthropology', *Annual Review of Anthropology*, 13: 447–66.
Sciama, D.L. (2003) *A Venetian Island: Environment, History, and Change in Burano*. Oxford: Berghahn Books.
Scheper-Hughes, N. and L. Wacquant (2002) *Commodifying Bodies*. Sage Publications.
Schneider, D. (1980 [1968]) *American Kinship: A Cultural Account*. Chicago: University of Chicago Press.
Schneider, D. and R.T. Smith (1973) *Class Differences and Sex Roles in American Kinship and Family Structure*. Englewood Cliffs, NJ: Prentice-Hall.
Schneider, J. (1971) 'Of Vigilance and Virgins', *Ethnology*, 9: 1–24.
Schneider, J. et al. (1972) 'Modernisation and Development: The Role of Regional Elites and Non-Corporate Groups in the European Mediterranean', *Comparative Studies Social History*, 14: 328–50.
Schneider, J. and P. Schneider (1973) 'Economic Dependency in the Failure of Comparatives in Western Sicily'. In Nash, J. et al. *Popular Participation in Social Change*. The Hague: Mouton.
―――― (1976) *Culture and Political Economy in Western Sicily*. London.
Segalen, M. (1991) *Fifteen Generations of Bretons*. Cambridge: Cambridge University Press.
Shore, C. (1992) 'Virgin Births and Sterile Debates: Anthropology and the New Reproductive Technologies', *Current Anthropology*, 33(4): 295–14.
―――― (1993) 'Ethnicity as Revolutionary Strategy: Communist Identity Construction in Italy'. In S. Macdonald *Inside European Identities: Ethnography in Western Europe*. Oxford: Berg.
Signorelli, A. (1996) *Antropologia Urbana*. Milano: Guerini Studio.
Slater, S. (1995) *The Lesbian Family Life Cycle*. London: Free Press.
Snow, C.P. (1998 [1959]) *The Two Cultures*. Cambridge: Cambridge University Press.
Spallone, P. (1989) *Beyond Conception: The New Politics of Reproduction*. Basingstoke: Macmillan Education.
Spallone, P. and L.D. Steinberg, (1987) *Made to Order: The Myth of Reproductive and Genetic Progress*. Oxford: Pergamon.
Spencer, J. (2000) 'British Social Anthropology: A Retrospective', *Annual Review of Anthropology*, 29: 1–24.
Spotts, F. and T. Wieser (1986) *Italy: A Difficult Democracy*. Cambridge: Cambridge University Press.
Stacey, M. (1992) *Changing Human Reproduction*. London: Sage Publications.
Stacul, J. (2003) *The Bounded Field: Localism and Local Identity in an Italian Alpine Valley*. Oxford: Berghahn Books.
Stanworth, M. (1987) *Reproductive Technologies*. Oxford: Polity Press.
Stolcke, V. (1986) 'The New Reproductive Technologies: The Same Old Fatherhood'. *Critique of Anthropology*, Amsterdam: Luna Publishers.
―――― (1988) 'New Reproductive Technologies: The Old Quest for Fatherhood', *Journal of International Feminist Analysis*, 1(1): 5–19.
Strathern, M. (1987) 'The Limits of Auto-Anthropology'. In A. Jackson *Anthropology at Home*. London: Tavistock Publication.
―――― (1990) 'Enterprising Kinship: Consumer Choice and the New Reproductive Technologies', *Cambridge Anthropology*, 14(1): 1–14.

——— (1991) 'Disparities of Embodiment', *Cambridge Anthropology*, 15(2): 25–43.
——— (1992a) *After Nature: English Kinship in the Late Twentieth Century.* Cambridge: Cambridge University Press.
——— (1992b) *Reproducing the Future: Anthropology, Kinship and the New Reproductive Technologies.* Manchester: Manchester University Press.
——— (1993) 'A Question of Context'. In J. Edwards et al. *Technologies of Procreation: Kinship in the Age of Assisted Conception,* Manchester University Press, Manchester.
——— (1998) 'Surrogates and Substitutes: New Practices for Old?' In J. Good and I. Velody *The Politics of Post-modernity.* Cambridge: Cambridge University Press.
——— (1999) *Property, Substance and Effect: Anthropological Essays on Persons and Things.* London: Athlone Press.
——— (2005) *Kinship, Law and the Unexpected: Relatives Are Always a Surprise.* Cambridge: Cambridge University Press.
Thompson, C. (2005) *Making Parents: The Ontological Choreography of Reproductive Technologies.* Cambridge, MA: MIT Press.
Valentini, C. (2004) *La Fecondazione Proibita.* Milano: Feltrinelli.
Vegetti-Finzi, S. (1998) *Volere un Figlio* Milano: Mondadori.
Warnock, M. (1985) *A Question of Life: The Warnock Report on Human Fertilisation and Embryology.* Oxford: Basil Blackwell.
Weeks, J. et al. (2001) *Same Sex Intimacies: Families of Choice and Other Life Experiments.* London: Routledge.
Weston, K. (1992) 'Forever Is a Long Time: Temporality and Authenticity in Gay Kinship Ideologies'. Unpublished paper.
——— (1997) *Families We Choose: Lesbians, Gays, Kinship.* New York: Columbia University Press.
White, C. (1980) *Patrons and Partisans: A Study of Politics in Two Southern Comuni.* New York: Cambridge University Press.
Williams, B. (1991) *The Politics of Culture.* Washington and London: The Smithsonian Institution Press.
Wolfram, S. (1987) *In-Laws and Out-Laws: Kinship a Marriage in England.* London: Croom Helm.
Xavier Inda, J. and R. Rosaldo (2002) *The Anthropology of Globalisation: A Reader.* Oxford Blackwell.
Yanagisako, S. (1978) 'Variance in American Kinship: Implications for Cultural Analysis', *American Ethnologist*, 5: 15–29.
——— (1991) 'Capital and Gendered Interest in Italian Family Firms'. In D. Kertzer and P. Saller *The Family in Italy: From Antiquity to the Present.* New Haven: Yale University Press.
——— (2002) *Culture and Capital: Producing Italian Family Capitalism.* Princeton University Press.

INDEX

A
Abu-Lughod, J. xxii
Adoption xi, xxi, 3, 6–7, 17, 30–31, 42–44, 52, 55, 60, 65–66,75, 77, 108–109, 113, 115
Anderson, B. xvi, 42, 95
Appadurai, A. xviii
Assisted conception x–xii, xiv–xix, xxii, 1–7, 15–19, 21, 25–32, 35–37, 52, 65, 67–70, 74, 79, 83, 95–96, 112–113, 116

B
Becker, G. xvii, 5, 23, 36, 40, 79
Bestard-Camps, J. 8, 10
Bharadwaj, A. xvii, 7
Biogenetics xi, xxi, 54–55, 57, 103–104, 106, 110–111
Biogenetic ties xxi, 102, 105, 110–111, 113, 115
Biological ties xxi, 5, 42, 53, 55, 65
Bonaccorso, M. xxii, 16, 32
Bouquet, M. 108
Bourdieu, P. 73, 83
Butler, J. 86

C
Cannell, F. 2
Carsten, J. xxi, 105, 108, 114
Catholic x, 12–14, 35, 39, 86, 119–125
Catholicism xix, 12–14, 35, 124
Choice 36–37, 41–42, 44–47, 49, 51, 55, 57, 64, 66–67, 70–71, 75, 81, 84, 89, 91–92, 94–95, 97–98, 105, 109
Clinicians x–xi, xiv–xv, xix–xx, 4, 7, 15–25, 29–33, 44, 47–48, 57, 63–64, 66–83, 96, 111–112, 118, 122, 126
Clinics xv, xix, 4–5, 7, 15, 17–20, 22, 25–26, 28–32, 35–36, 44, 47–48, 60–62, 67–70, 73, 80, 83, 85, 95–96, 112, 122, 126
Commonplace(s) xx, 43–44, 67, 79–80
Corea, G. 3
Cussins, C. 5, 23, 44, 48

D
D'Andrade, R. and C. Strauss 114
Daniels, K. 57, 65
Daniels, K. and E. Haimes xvii, 4, 57, 64
Davis-Floyd, R. and J. Dumit xvii, 35
Delaney, C. and S. Yanagisako 92
De Pina-Cabral, J 8, 10, 12, 14
Dolgin, J. xvii, 4, 5, 109, 114
Donor(s) xv, xx–xxii, 4–6, 17, 19, 23, 26, 32–34, 40, 43, 45–46, 51–54, 56–64, 66–69, 72–73, 76, 79–82, 85, 06–97–105, 111–113, 115, 118–119

E
Edwards, J. ii, xiii, xvii, xviii, 3, 10, 28, 56, 59, 62, 114
Edwards, J. and M. Strathern 114

Egg donation xv, xxii, 23, 40–41, 51–52, 71
Embryos xi, 59, 61–62, 69, 78, 122, 124–125
Eterologa 23, 40–43, 46–47, 58, 71, 119
Ethnography xvi, xix–xxii, 4, 15, 17–18, 31, 85, 107
Euro-American xvii, xix, xxi, 1–3, 63, 107, 111, 113–114, 116

F
Fabian, J. xxii
Family xi, xv–xvi, xix–xx, 2–6, 8–11, 34–35, 37, 39–43, 54–56, 64–65, 77, 81–85, 89–94, 96–97, 101, 103–106, 108–116
Finkler, K. 76
Franklin, S. xvii, 3, 4, 5, 10, 23, 41, 47, 50, 108
Franklin, S. and M, McNeil 3
Franklin, S. and H. Ragoné xvii, 5
Franklin, S. and S. Mckinnon 108
Frith, L. 64

G
Galt, A. 8
Gambetta, D. 9
Gamete donation x, xiv–xvi, xix, xx–xxii, 3–4, 6–7, 15–19, 21–24, 28–32, 34–37, 40–49, 52–53, 55, 57, 61–62, 64–73, 75–77, 80–83, 85, 88, 96, 99, 102, 106–113, 115
Gell, A. 10
Gift(s) 59, 62–63, 73
Green, S. 25
Gunning, J. and V. English 69
Gupta, A. and A. Ferguson 17

H
Haimes, E. 4, 40, 64
Hall, S. 126
Handwerker, L. 7
Hannerz, U. xviii
Hastrup, K. and K. Olwig 17
Hayden, C. 91, 94, 95, 99, 100
Heterosexual xiv–xvi, xix, xxii, 15–18, 22, 26–31, 33, 35, 51, 66,
85–89, 91, 93–96, 99, 101–102, 105–110, 113, 115
Hirsh, E. 3
Hobsbawm, E. and T. Ranger 41

I
Identity xi, 9, 28, 32, 41, 55, 57, 86, 110, 116, 119
Infertility xi, xvi, xix, 4–7, 15–16, 18–19, 21, 24, 28–30, 34–40, 42–44, 46, 48, 50, 52, 56–57, 59–60, 62, 67–71, 75–78, 80–81, 85, 96, 107–112, 116
Inhorn, M. ii, xviii, 6
Inhorn, M. and F. Van Balen xviii, 6
Italian Kinship xv–xix, xxi, 1–2, 9, 113
Italy x–xii, xiv, xvii–xviii, xix, xxi–xxii, 1, 6–10, 12–15, 17–18, 26, 31, 35, 46, 51, 68, 70, 86, 91, 94–95, 112, 115, 117–118, 123, 126
IVF 4, 6, 17, 19, 23, 29–31, 45, 47–51, 57, 59, 77, 81–82

J
Jackson, A. 16

K
Kahn, S. xviii, 6, 108
Kinship xi–xii, xv–xxii, 1–11, 13, 16–18, 31, 67, 81–85, 103–105, 107–110, 113–115
Konrad, M. ii, 4, 45, 59, 63, 64, 81

L
Lasker, J. 57, 58, 65
Lay couples xiv, 28
Law 68, 78, 112, 117–126
Lesbian and gay couples xiv–xv, xvi–xx, 16–18, 25, 27, 31, 45, 57, 69, 85–91, 94–96, 98–102, 104–106, 109–111, 113, 115–116
Lewin, E. 89, 90, 94, 100

M
McNeil, M., I. Varcoe, and S. Yearley 3
Marcus, G. xviii, 17

Marcus, G. and M. Fischer xxi
Medical language xx, 40, 67, 79–81
Mediterranean x, xvii, xix, 8, 10–12
Miller, D 8, 71
Model x, xvi–xviii, 2, 39, 45, 87–89, 105, 107, 111–112, 113–116
Mol, A. 76
Morely, D. 126

N
Nielsen, L 35, 69
Novaes Bateman, S. 5, 59, 64, 69, 75

O
Okely, J 16
Ong, A. and S. Collier 16
Orobitg, G. and C. Salazar 6
Ovo-donazione 40, 71

P
Pardo, I. 9
Pies, C. 90
Plesset, S. 9
Pregnancy xi, 34, 41–42, 48, 58–59, 62–64, 73, 105
Price, F. xvii, 3, 4, 69, 72
Procreation xv, 3, 42, 73, 85, 108, 111, 113–116, 120

R
Rabinow, P. 73
Ragoné, H. xvii, 4, 5, 10, 23, 35, 41, 42, 47, 70, 71, 81, 137, 139
Rapp, R. 73
Relatedness xv–xvii, xx, 5, 42, 46, 82, 85, 105, 108, 112–116
Reproduction Xii, 2, 27, 78, 108–109, 112, 124, 126
Rhetoric(al) xi–xii, xx, 5, 59, 67, 70–71, 81, 88, 93, 112
Rivière, P. 2
Rowland, R. 79

S
Saffron, L. 89, 90, 97, 99, 100
Saunders, G. 8, 9, 15
Scheper-Hughes, N. and L. 60
Schneider, D. xvii, xx–xxi, 9, 11, 53, 99, 107–110, 114–116
Sciama, D. L. 9
Segalen, M. 8
Shore, C. 2, 13, 14, 16
Signorelli, A. 8
Slater, S. 88
Snow, C.P. 19
Social ties xvi, 6, 9, 42, 104, 106, 110–111, 113
Southern Europe(an) xvii, xix, 107, 114, 116
Spallone, P. 3
Spallone, P. and L. D. Steinberg 3
Spencer, J. 10
Sperm donation xv, xxii, 4, 17, 19, 40, 60, 76, 118
Stacey, M. 3
Stanworth, M. 3
Stolcke, V. 40, 70
Strathern, M. xiii, xvi–xvii, xxi, 2–3, 9, 16, 20, 46, 57, 62–65, 70, 76, 107, 109–110, 114–116

T
Technology xii–xiii, 3–4, 6, 51, 57, 70–71 77, 79, 108
Thompson, C. 75. *See also* Cussins

V
Vegetti–Finzi, S. 6

W
Warnock, M. 2, 46, 65
Weeks, J. 87, 89, 90, 100
Weston, K. 87, 89, 91, 94, 99, 100, 103, 104, 105, 109, 111
Williams, B. 113
Wolfram, S. xx, 9, 53, 114

X
Xavier Inda, J. and R. Rosaldo xvii

Y
Yanagisako, S. 9, 10, 92, 108, 114